Irish Nature

for Sean and Mark
my grandsons

First published 1980
The O'Brien Press Ltd.,
11 Clare Street, Dublin 2, Ireland

© Copyright reserved
ISBN 0 905140 39 7

Jacket design: Jarleth Hayes
Binding: John F. Newman Ltd.
Typesetting and Layout: Redsetter Ltd.
Printed in the Republic of Ireland
by A. Folens & Co. Ltd.

Irish Nature

NORMAN HICKIN

THE O'BRIEN PRESS
11 CLARE ST DUBLIN 2
11 CLARE ST DUBLIN 2
THE O'BRIEN PRESS

Contents

Part One
The Animal Kingdom

Part Two
The Vegetable Kingdom

Acknowledgements

In a work of this description it is obvious that I have received help and support from many people. Indeed, it has been a source of great encouragement to me to experience the unstinting assistance and co-operation of so many people with specialized knowledge of Ireland's fauna and flora. Many authorities have read the different sections for me and have made valuable comments and suggestions but I hasten to add that any errors of fact or degree remaining in the work are entirely my own responsibility. A number of fellow naturalists have lent specimens to me for description and drawing, or have informed me of their whereabouts. Specialists have identified my own specimens collected in the field or have sent me literature on their subjects. I extend my gratitude to all. It is difficult to list all who have helped in these ways but I feel that I must make special mention of those whose names follow.

Firstly, I would like to give my thanks to my publisher, Michael O'Brien, whose enthusiasm for this book seldom flagged. A visit to him always stimulated and refreshed me. We are still firm friends! Then I would like to thank especially Dr. D. Naylor and my daughter Verney, his wife, for their help in allowing me the use of their home in Dublin as a base for my visits to the National Museum and other Dublin-based institutions, including my publisher's office. In addition, their cottage in West Cork has served as a valuable work-station and I am deeply grateful to them both. I would also like to thank Dr. J. Bracken, Dr. C. O'Riordan, Mandy Precious, Freda Adair, Dr. A. Cook, Dr. J. P. Cullinane, Dr. M. Telford, Dr. C. Moriarty, Maura Scannell, R. McMullen, Dr. P. Merrett, Dr. P. Pitkin, Dr. M. Speight, J. M. C. Holmes, A. Barber, D. Synnott, Veronica Burns, P. Whalley, N. Sullivan, Professor J. L. Cloudsley-Thompson, Patricia Greacen, P. Goyvaerts, B. F. Skinner, Cynthia Longfield, Dr. J. P. O'Connor, D. Street, O. Merne.

Pamela Guilder-Willis, my Personal Assistant, has not only been responsible for the large volume of correspondence which the project has engendered but has typed all the manuscripts. This has entailed deciphering my handwriting and correction of grammar. I am deeply grateful to her for the great contribution she has made to the book.

I am grateful to a number of national establishments whose staff have given help in the identification of specimens, the loan of material for drawing and for numerous facilities. I earnestly hope that they feel that their efforts here proved worthwhile. The National Museum of Ireland; the British Museum (Natural History); Irish Biological Records Centre; Trinity College Department of Geology, Dublin; Irish Biogeographical Society; Forest and Wildlife Service of the Department of Lands; National Botanic Gardens, Glasnevin; University College Dublin; New University of Ulster, Coleraine; Institute of Terrestrial Ecology, Dorset, England.

Lastly to Emma, my wife. I owe an immense debt of gratitude for her constant encouragement over the years, especially when the work load was heavy, prolonged or unrewarding. Additionally, Emma has tolerated the litter of books and specimens in our living quarters and has even allowed road-casualty specimens to be stored in the household fridge — *almost* without demur.

Introduction

This book describes and illustrates the broad spectrum of nature in Ireland — both plants and animals. In this introduction I shall describe some of the geographic, climatic and human factors which have influenced the unique combination of plants and animals in Ireland.

Ireland is a relatively small island, a little less than five hundred km long from the north-east to the south-west, and a little less than three hundred km wide from west to east. In contrast with the east coast, which faces the Irish sea, the west coast, facing the wild Atlantic Ocean, is considerably indented. Ireland's total area is about 83,000 km.

Ireland's position in the Atlantic Ocean is very important to its climate. This great ocean has provided a moderating influence. The temperature of the sea around Ireland is higher than that of the land in winter, whereas in summer the land heats up rapidly but is constantly tempered by the relative coolness of the Atlantic. The comparative warmth of the water is caused by its movement from equatorial latitudes, and this movement is known as the North Atlantic Drift. Its effects were noticed many centuries ago when early Scandinavian people observed strange seeds and other vegetation washed up on their shores, which they surmised must have come from far away places. Such material finds its way to the Irish coastline today. The Scandinavian people call this movement of water the *Strøm,* which is the derivation of the present popular name 'Gulf Stream'.

Ireland lies between latitudes 51° and 56°N, which puts it in the path of westerly air movement. These prevailing winds are saturated with moisture from the Atlantic, and when the moist air meets the cooler land, the result is precipitation, either rain or a fine drizzle. Substantial amounts of snow are rare. The mountainous areas of the west, and to a lesser extent of the east, receive rainfall of approximately 1,600mm annually. Prolonged periods of freezing weather are rare. In summer, however, temperatures are generally lower than they are in continental Europe, and long periods of hot weather bringing droughts are unusual.

Since human beings arrived in Ireland, over 8,000 years ago, they have caused profound changes. At that time, Ireland was covered with extensive forests, but when the change was made from hunting and scavenging to tilling the soil, large forests began to disappear. The land they occupied was cleared for agriculture. During historical times this process continued with the felling of trees for fuel and for the manufacture of charcoal. With the development of commerce and with the beginnings of sea-borne trade, new animals and plants were introduced to the country, some of which have become pests. More recently, the replanting of de-forested areas with commercial trees not previously found in Ireland has again changed the landscape.

For a recent account of the physical environment (including the geology and an account of the original composition of Ireland's flora) the reader is directed to the first two chapters by Mary Davies and E. C. Nelson in the book, *Irish Gardening and Horticulture,* published by the Royal Horticultural Society of Ireland in 1979.

It is undeniable that most of the natural history books consulted in Ireland deal with British natural history. A small section may be devoted to the Irish element — often noting only the presence or absence of Irish plants and animals. This book, however, deals exclusively with Irish wildlife.

It is somewhat daunting to try to describe all Irish wildlife within the covers of a single volume, but this I have attempted to do. It was never my intention, however, to produce a scientific textbook filled with keys for identification, using large numbers of scientific terms (though I do not mean to decry such books). The present work is intended to give the general reader an appreciation of the animals and plants in the Irish environment. I have described species that should be conserved and cherished, and I have also included some 'pest' species which, in the interests of wildlife management or of human hygiene, must be controlled. The task of choosing representative species was difficult. In some Groups, almost every species has been illustrated and accompanied by notes. This is true for the mammals, but for other Groups it was an impossible task. In some cases I chose to illustrate a species I had encountered in its natural habitat in favour of one I had not seen. Almost without exception, drawings were made from Irish material, either from the Museum collections, or from specially collected specimens.

Classification is, of course, very important to this book, but it has been dealt with only in a descriptive way and only where absolutely necessary. Professor O'Rourke has covered the classification of animals in his *Fauna of Ireland,* and Irish plants have been classified in Dr Webb's *An Irish Flora.*

In this review of Irish wildlife I tried to include any species with unique traits. One example is the Hiberno-Lusitanian wildlife. A number of animals and plants found in the south-west of the country are to be found elsewhere only in the western extremities of France, Spain, Portugal, and occasionally in the south-western tip of England and Wales. In addition, some animals and plants, well known in Europe generally, show features which differentiate them from the European stock. Although differences are often minor, they are usually constant and are given sub-specific rank. Since the land-masses of Ireland and Britain separated, the fauna and flora of Ireland have existed in comparative isolation, and this has caused changes which account for these minor variations. Naturally, I have also discussed some of the more common species for the general reader who may wish to know more about the animals and plants he sees about him.

For the illustrations I have used what is called the 'scraperboard' method. This technique entails drawing on a layer of prepared chalk on a board backing. The outline is drawn and then filled in by painting over with Indian ink; this is the silhouette stage. When the drawing is dry, excess ink is removed by scraping with special tools. In several cases, I have combined traditional line drawing with the scraperboard technique, using both a mapping pen and scraperboard instruments.

There are a number of advantages in using this technique for wildlife drawings. It allows the principal shape and the important detailing (for identification of the animal or plant) to be emphasized in a dramatic contrast of white against black. This contrast has high impact value on the viewer, and drawings that employ this method may thus be more memorable than traditional line drawings.

In nature, few surfaces are hard or devoid of detail. Nearly all are hairy, bristly, covered with tubercles, feathery, scaly or slimy, and many forms of microscopic detail cannot be portrayed in a drawing. With scraperboard, however, the illustrator may use a number of methods to show these features.

The book is divided into two main sections, The Animal Kingdom and The Vegetable Kingdom. There is a main index and in addition a geographical index of all the places mentioned. All the main entries are listed separately in Latin, English and where possible in Irish.

I hope the reader will enjoy looking at this book and also that it will be a useful reference source for all who take an interest in the wonders of nature.

Norman Hickin 1980

PART ONE
The Animal Kingdom

1 Mammals of the Land

Hedgehog, Shrew, Bats, Rabbit, Hares, Rodents, Carnivores, Deer, Wild Goat, Pony

MAMMALS *Mammalia*

Mammals constitute the most highly developed class of all animals, but show great variation in form and size. They include such diverse forms as whales, bats, mice and man himself, but all have essential characteristics in common. The most important of these is the possession by the female of highly modified cutaneous glands, similar to sweat glands, which secrete milk for feeding the young after their birth. These glands are known as the mammary glands. Initially, the milk forms the only nourishment taken by the newly-born offspring, but later other foods are taken also until the young animal is independent of its mother's milk (the time of 'weaning'), and it is able to fend for itself. The timing, amount and the duration of the flow of milk is controlled by hormones.

Associated with this character is the possession by mammals of a clothing of hair. Sweat glands are modified follicles in the skin from which a hair grows. There is a great variation in the amount of hair on the body, and whereas some mammals may be more or less completely covered with hair, others, such as whales, have only a few hairs arising from the throat region which are soon lost after birth. Two or more kinds of hair are present on some mammals. There may be a dense, under-fur protected by long, bristly, guard-hairs, and also long, sensitive 'whiskers' on the muzzle and face, and these are well-supplied with nerves at the base. In this way much information is collected by these whiskers which is very important, especially in nocturnal mammals and those swimming in muddy water or at great depths.

The prickles of the hedgehog are modified hairs, and this mammal possesses fur-like hair also, on the face, legs and belly, as well as 'whiskers'. All mammals are warm-blooded. That is to say that the blood is maintained at a constant, elevated temperature — usually about 36°C (98°F). This allows the mammal to sustain a high degree of activity, even in adverse conditions of temperature, although a number of small mammals pass through periods of suspended activity (such as hibernation) in severe cold or aestivation in high temperatures, and the covering of hair together with the sweat glands help to regulate body temperature. It should be noted that birds also are warm-blooded.

Mammals show a number of anatomical features which differentiate them from other classes of animal. Chief among these is a four-chambered heart. Birds and crocodiles share this feature but it has been acquired along different evolutionary routes. Usually, but not invariably so, the brain is larger than in other classes, allowing a higher degree of intellectual versatility. Mammals possess only seven neck (cervical) vertebrae. Even the giraffe has only seven but the number present in birds and reptiles is subject to extreme variation.

Of the eighteen orders of living mammals occurring in various parts of the world today, members of nine are found in Ireland, either truly wild or in a feral condition. These are given below together with the number of species in each.

Insectivora	Hedgehog, Pygmy shrew	2 species
Chiroptera	Bats	7 species
Lagomorpha	Rabbit, hares	3 species
Rodentia	Rats, mice, squirrels, vole	7 species
Cetacea	Dolphins, whales, porpoises	22 species
Carnivora	Fox, marten, stoat, mink, badger, otter	6 species
Pinnipedia	Seals	2 species
Perissodactyla	Pony	1 species
Artiodactyla	Deer, goat	4 species

Insectivora

Members of the *insectivora* are generally small in size and, although appearing very different from each other (as for instance shown by Ireland's only two species, the hedgehog and the Pygmy shrew), they share many common features. These are usually of a primitive nature. There is always a mobile 'muzzle' or snout projecting in front of the teeth which is often provided with numerous, sensitive whiskers. The small eyes and ears are usually more or less hidden in the fur and sometimes in the skin. Sex determination from external features is frequently a problem because the urinogenital system has only a single, external opening. The skull is generally narrow, low and flat, and the braincase is usually small and does not bulge upwards. The small brain is simple in structure and although the olfactory lobes are large, the cerebral hemispheres are non-convoluted and smooth. The numerous teeth are non-specialised, and in each half of each jaw there are at least two incisors. Although the name of the group implies an exclusive insect diet, worms and molluscs are often eaten, as well as carrion from time to time, and vegetable matter.

Hedgehog *Erinaceus europaeus*

The hedgehog is one of the best known of all our wild mammals, especially to the country dweller. Its habit of rolling into a ball and extending its prickles in all directions when threatened is known to everyone. A detailed description is, therefore, unnecessary. Although the hedgehog population underwent a decline at the start of this century and was said to be scarce in a number of places in 1935, it is now widespread in areas suitable for it. With the large number of road casualties, it is surprising that the hedgehog numbers are maintained at all, and in Britain these appear to make little difference in suburban areas where the hedgehog is well established and abundant. Surprisingly, they are to be found in many gardens and parks wherever there are little-disturbed areas in which they may spend the day, nest and hibernate in piles of dead leaves and garden rubbish. Generally, however, they occur in wooded areas, broadleaved trees being preferred to coniferous, which are not too dense, or in more open country where there are hedges or other suitable cover. Life for the hedgehog involves being abroad a short time before dusk and through until the first glimmer of dawn, but his presence is recorded by the total disappearance of a plate of kitchen scraps left at the door and by the tell-tale black, rounded droppings (often decorated with beetle elytra) on the lawn. For those walking in the evening the hedgehog makes his presence known by the snuffling noises or quick patterings along the lane verges.

Hedgehog, *Erinaceus europaeus*
Male about 230mm, female about 220 mm
in length. Widespread and generally distributed.

The hedgehog's nest is always well concealed. Deep in a pile of vegetable debris lies a ball of mixed, dried grass and leaves with usually some protection from the climatic conditions by being sited near a few bushes. The pregnant females are found from May until October and two litters may be produced annually. Gestation takes 31 or 32 days, and the young are born blind and with soft prickles. They open their eyes in about 14 days, and leave the nest soon after 21 days. They present a happy sight following their mother in single file! The main food consists of insects, slugs, earthworms, snails and millipedes, but young mice and birds, as well as a small amount of vegetable matter in the form of berries, are also taken.

Pygmy shrew *Sorex minutus*
The shrews are amongst the smallest mammals and although the Pygmy shrew is the smallest Irish mammal, another species, Savi's Pygmy shrew, *Suncus etruscus,* from the Mediterranean region, is the smallest known mammal. The Pygmy shrew is widespread throughout Ireland, even to the tops of the highest mountains, and is found virtually in any type of habitat where there is either little or some vegetative ground cover into which it may burrow, but where it is unable to use the tunnels of other mammals, such as those of the Wood mouse.

The Pygmy shrew's muzzle is long, extending well forwards and appears to be somewhat swollen where the long, sensitive 'whiskers' (the vibrissae, up to 16mm in length) are inserted. Said to be 'sandy-brown' in colour but appearing to be much more greyish or nearly black with a velvety sheen, it has a densely-haired tail which is also thick and long. It weighs about 5gm[1]

It is extremely active and emits many squeaks and twitters, reacting swiftly when confronting other small mammals. It often enters buildings, including homes, and will climb to the upper stories, although where it obtains the large food intake required for this time and energy-consuming feat, is a mystery. The high point of breeding is about April or May but it seems that the Pygmy shrews which are born late in the year do not become sexually mature until the spring of the following year. Gestation is about 22 days and the lactation period is about the same. There are probably two litters per season. The Pygmy shrew feeds on small invertebrates, woodlice, spiders and beetles being favoured.

Bats *Chiroptera*
Bats are readily distinguished from other mammals by their capability of maintained flight. This is unique among mammals. Although they are comparatively little known, about 800 world species have been described. The fore-limbs show exceptional modification, with the strong, upper arm relatively short and the fore-arm long and consisting of only a single bone. The wrist is compact with some of the bones fused together, but the thumb is free and projects forwards so that its claw may be used for climbing. The fingers are very long, supporting the thin wing membrane. The latter consists of two layers of skin which is continuously joined to the fingers and thence to the ankle of the back leg. It also stretches between the back legs, taking in some part or all of the tail. Tendons are moved by muscles in the arm which can give separate movements to the various regions of the 'wing', allowing exceptional mobility. Indeed, it is said that bats are by far the most manoeuver-

Pygmy Shrew, *Sorex minutus.* Length of head and body 58 mm, tail 38 mm. Widespread and abundant.

Lesser Horseshoe Bat, *Rhinolophus hipposideros*. Length of head and body 35 mm. Common locally in western seaboard Counties.

able of aerial mammals. The hind feet of bats are short and round and the five toes bear sharp claws by means of which the bat is able to hang, head downwards, in its resting place.

Bats show extraordinary breeding behaviour. They are entirely promiscuous and pair formation is unknown in this group of mammals. Mating takes place in the autumn, but the sperm is internally stored by the female and it is not until the following spring, after hibernation, that fertilization occurs. The young are born well-developed and in all-female colonies. The young bat is at first carried by the mother during her insect-hunting flights. When becoming too heavy for this, it remains in the colony roost.

All bats are nocturnal, but their eyes are generally small and are not relied upon for identifying and hunting prey. This is accomplished by a type of echo-location, and when the bat is active a continuous stream of sound pulses is emitted. The source of these sounds is a modified voice-box and the duration of the pulses may vary from as little as one-quarter of a millisecond to about 60 milliseconds. They are almost entirely inaudible to man. The bat receives the echo through its ears, which are usually large and sometimes exceptionally large. However, in some species there is a separate piece (the earlet or tragus) situated in front of, and at the base of, the main earlobe. That part of the brain associated with hearing is extremely large. All bats are insectivorous, but in other parts of the world, a wide range of food material is taken. No bats make nests. The seven species of bats in Ireland are as follows:

Lesser Horse-shoe bat	*Rhinoclophus hipposideros*
Leisler's bat	*Nyctalus leisleri*
Long-eared bat	*Plecotus auritus*
Pipistrelle bat	*Pipistrellus pipistrellus*
Natterer's bat	*Myotis nattereri*
Whiskered bat	*Myotis mystacinus*
Daubenton's bat	*Myotis daubentonl*

Lesser Horse-Shoe Bat
Rhincophus hipposideros

This is the sole Irish member of the family *Rhinolophidae*. With a wing-span of up to 250mm, this species is greyish-brown in colour with the underside somewhat paler. It weighs about 6gm in December but decreases to less than 5gm by April. It possesses strange, leaf-like organs of skin on the nose and there is a large, horse-shoe shaped piece with a finger-like protrusion pointing upwards from the centre, known as the sella. This in turn is surmounted by the lancet which is itself nose-shaped with another finger-like piece also pointing upwards. The eyes are situated close together at the base of the lancet. At the base of the ear there is a horizontal, shelf-like organ known as the anti-tragus. The Lesser Horse-shoe bat can produce a high-pitched squeak. Living in small colonies, it rests with its wings wrapped around its body and hangs quite freely from a ledge, whereas all other bats prefer to hang against a vertical surface. It takes its prey to a convenient perch where it is divided up before being swallowed. It is common locally in Clare, Cork, Galway, Kerry, Limerick and South Mayo.

Whiskered Bat *Myotis mystacinus*

This is another species which is common almost throughout Europe and Asia and it is almost certain that it is generally plentiful in Ireland but, being a bat which is not usually recognized, it often goes unrecorded. The long, silky hair of the whiskered bat is brown but the tips are sometimes pale. The underside is white. Its wing span varies from 225 to 245mm and its weight from 4.5 to 6gm. The flight is rather slow and fluttering and is usually at a height of 1 to 5m, and it squeaks in a low, buzzing manner. The small insects and spiders on which this bat subsists are generally taken from hedges and fences, but some are flying insects and taken on the wing. The Whiskered bat may be differentiated from others likely to be found in Ireland by a bony extension of the hind foot (the calcar), which traverses the margin of the interfemoral membrane at least halfway to the tail.

Natterer's Bat *Myotis nattereri*

Three of Ireland's bats belong to the genus *Myotis*. Natterer's bat has been recorded from many localities in Ireland, and is found in most of Europe, in Asia and as far as Japan. It is a medium, greyish-brown on top and the underside is white. From the ear to the shoulder the demarcation line is sharply defined. The wingspan is from 265 to 285mm and the weight of an adult may vary from between 8 to 9.5gm. During flight the tail is directed downwards at an angle of $60°$ to the body. Its definitive characteristic is a fringe of hair on the interfemoral membrane on each side of the tail. It generally feeds on moths, some of which are plucked from their resting places.

Daubenton's Bat *Myotis daubentoni*

Daubenton's bat is yet another species of wide distribution in Europe and Asia. It is already known from eleven Irish counties and almost certainly will be found to be even more widespread. On its back the fur is medium to dark brown with paler tips, but the underside is greyish and there may be a yellowish tint. The wing span is up to 245mm and the weight may vary between 7 and 11gm. This species is positively identified by the length of the calcar (see Whiskered bat) which extends two-thirds along the outer margin of the interfemoral membrane. Although usually silent in flight, when frightened it will 'buzz' angrily.

Leisler's Bat *Nyctalus leisleri*

Leisler's bat is probably the most common and widely distributed bat in Ireland, whereas in Britain it is considered to be rare. On the other hand, the Noctule bat, *Nyctalus noctula,* which is closely related but a little larger, is common in England and Wales but apparently absent from Ireland. Its wing-span is up to 320mm and it weighs between 15gm. and 20gm. The fur is dark brown on the back and slightly paler underneath. There is no nose-leaf present, as with all other Irish bats, but there is a skin lobe at the base of the ear, at the front, known as the tragus. This has a special shape, being widest at the top and it is this characteristic which enables Leisler's bat to be distinguished from all other Irish bats. It utters a rather metallic, high-pitched squeak, but screeches at other bats. Usually flying at a height of 3 to 17m, it makes shallow dives.

Pipistrelle Bat *Pipistrellus pipistrellus*

The Pipistrelle is generally recognized as a most abundant bat, but also a common bat around buildings, where colonies of several hundred individuals are not unusual. It is widely spread throughout Europe. It is the smallest Irish species. The fur on the back is medium to dark brown in colour and may have a reddish hue. The underside is lighter. The wing span is up to 230mm but the weight is only from 3 – 7.5gm. and it will creep into the most confined spaces. Although the period of its hibernation lasts from the end of October to early March, it flies frequently in winter whenever the temperature rises above $4.5°$C. Flying at medium height but often reaching 13m, it has a definite beat which it traverses in a rapid but jerky flight. Its main food is said to be gnats, but larger insects are taken to a perch to be dealt with.

Long-eared Bat *Plecotus auritus*

This is one of the most abundant Irish bats. Weighing from 5 gm. to 9.5gm. this species has a wing-span of up to 235mm. It is easy to identify by its exceptionally long ears which are almost as long as the head and body together. In flight these are held erect and slightly forwards but can be tucked away when the bat is at rest. When being unfolded they have a ram's horn appearance. The fur on the back is brown to greyish-brown with the underside yellowish-brown to yellowish-white. The long-eared bat often flies during mild weather in mid-winter, and migrations probably occur. It is known to feed on hibernating, Small tortoiseshell butterflies in buildings, and this is likely to be the explanation for several hundred wings of this butterfly being found in a building near Bantry at Christmas after a two months' absence. Resting prey is usually taken and large insects are taken to a perch for cutting up.

The Rabbit and Hares *Lagomorpha*

The rabbit and hares, the lagomorphs, were formerly included with the rodents to form the order *Rodentia*. This was on account of the possession of a diastema (the wide gap between the incisor teeth and the premolars). The incisors are adapted for gnawing by growing continuously and thus always forming new cutting edges. However, they are now considered to be two separate orders mainly because the lagomorphs possess two pairs of upper incisors, as well as a pair of non-functional ones which are situated immediately behind them.

The family LEPORIDAE, with which we are solely concerned in the Irish fauna, consists of the Common rabbit, *Oryctolagus cuniculus* and the two species of hare, the Brown hare, *Lepus capensis,* and the Irish hare, *Lepus timidus hibernicus*. All three are characterized by being medium-sized animals with long, narrow ears and short tails. The forefeet have five toes and the hind feet four, and all have sharp claws. The hind legs are considerably longer and stronger than the forelegs and all feet are well supplied with a brush of tough hairs, the latter giving a non-slip effect when running. The subdued fur colour of brownish tones with black speckling

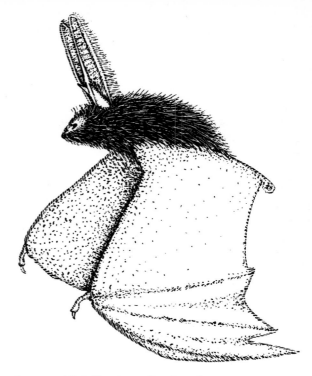

Long-eared Bat, *Plecotus auritus*. Length of head and body 46 mm. Generally distributed and often abundant.

or small patches of white, makes them inconspicuous, but if menaced by a predator they seek escape by swift flight. The three species are strictly vegetarian and food is twice passed through the body. This is known as coprophagy or refection.

Rabbit *Oryctolagus cuniculus*

The rabbit is a most widely distributed animal, being found throughout Europe and North Africa. In addition, it has been introduced into Australia, New Zealand, the USA and Chile. It is most abundant in open grassland, cultivated land generally, and open deciduous woodland, but it is often present in considerable numbers in coastal areas where there are sand-dunes and saltmarsh. Other habitats where it occurs are mountainous regions, moorland and sea-cliffs. It generally lives in a burrow or warren which may be a simple or complicated tunnel system, but in recent years it has tended not to go underground but to use thick scrub or hedge debris. During the last twenty years or so, rabbit populations have been periodically, and often drastically, reduced by Myxomatosis.

Their colour is usually buff or buffy-grey with black speckles, but the nape is reddish. The underparts are white whilst the tail is black above and white below and this makes it conspicuous when the animal is running away. However, many colour variations, including black, often occur. Breeding takes place principally in the first six months of the year, but sporadically at other times. A nest made from grass or straw with an inner layer of fur plucked from her body is constructed by the doe. Gestation is 28 days and the first litters are small in number but later ones large, the numbers varying from 3 to 7. Embryo litters may be partially or wholly absorbed. Lactation lasts about 3 weeks and the young rabbits become sexually mature in three to four months but go on increasing in size up to about nine months.

The rabbit eats about a half-kilo of fresh green food daily, but its attraction for agricultural crops makes it an injurious species of great importance.

Brown Hare

Lepus capensis formerly *europaeus*
The Brown hare or European hare has been introduced into Ireland where it flourishes in the north-west. It is very widely distributed, being found throughout Europe, Asia and Africa and has been introduced into Australia, New Zealand, South America and the Great Lakes area of the United States of America.

The back hair is tipped with black giving a rough appearance. The ears are long, with acute apices which are black tipped. The tail has a black upper surface whilst the lower is white, but it is tucked underneath between the legs when running, so that no white is showing as in the rabbit. In late summer or early autumn it moults to give a dense, reddish-coloured coat. It has a straw-coloured iris. It weighs up to 4kg. When wounded, or in terror, it screams, but its swift running and endurance enables it to out-run most predators. It rests in 'forms' in the open, where it squats close to the ground with the flattened herbage underneath. At the be-

Brown Hare, *Lepus capensis*. 545 mm in length. Introduced into Co. Donegal where it does well.

ginning of the breeding season hares, which are otherwise solitary, consort together and there is much chasing, leaping and sparring. Gestation is 42 to 44 days with a second conception a few days before birth, so that two sets of foetuses are present simultaneously. Many embryos, however, die and are subsequently absorbed. The number in the litter varies from one to four, and the young are born completely haired and with eyes open. No nest is made, the young being born directly into the form. Sexual maturity is not reached in the year of birth, generally taking about eight months. There are three or four litters per year.

Irish Hare *Lepus timidus hibernicus*

This is the Irish race of the Mountain or Blue hare, the various races of which occur throughout the temperate areas of the Old World. It is much less specialized than is the Brown Hare, eating a wider variety of plants and also occupying areas from sea-level to high altitudes. Often it gathers into small groups. At one time, large numbers could be seen by aircraft passengers when landing at Belfast Airport, as a large 'herd' grazed the grass between the runways. The tail is partially white on the upper surface. Its coat in summer is a dusky-brown with the tips of the

Irish Hare, *Lepus timidus hibernicus.*
545 mm in length. Abundant
throughout Ireland.

hairs grey – giving it a blue appearance but
frequently the coat is russet or even foxy-red,
usually with some whitening. It undergoes three
moults and, whereas the Scottish race is white
in winter, only a proportion of the Irish race
attain a white coat during this season. The breeding season is longer than that of the Brown hare,
with the litter having a maximum of three young.
There may be up to three litters annually.

Rodents *Rodentia*

There are more species of rodents than of any
other mammalian order – about 1,500 species
being known. They have adapted to almost every
type of terrestrial habitat and they are almost all
small, rat, mouse or squirrel-like. Rodents are
easily distinguished from all other mammals by
their teeth. Only one pair of incisors in both
upper and lower jaw is present, and they grow
continuously throughout life. At the front of the
tooth there is a thick coating of hard enamel
which forms a chisel-like edge as the soft dentine
at the back wears away much more quickly.
There is a gap between the incisors and the
cheek teeth, known as the diastema, and the
lower jaw is only loosely articulated, which
facilitates the gnawing action. Predominantly
eaters of hard vegetable matter such as seeds,
rodents are, to a greater or lesser degree, omnivorous. In the Irish fauna are included three cos-
mopolitan and true commensals of man – the
House Mouse, *Mus musculus;* the Brown Rat,
Rattus norvegicus and the Black or Ship Rat,
Rattus rattus.

Irish Rodents

Red Squirrel	*Sciurus vulgaris*
Grey Squirrel	*Sciurus carolinensis*
Field Mouse	*Apodemus sylvaticus*
Black Rat	*Rattus rattus*
Brown Rat	*Rattus norvegicus*
House Mouse	*Mus musculus*
Bank Vole	*Clethrionomys glareolus*

Red Squirrel *Sciurus vulgaris*

This attractive squirrel is distributed in woodland areas over a large part of temperate Europe
and Asia, from Ireland to Japan, although it
exists in many colour variants. In alpine Switzer-
land they are mostly black and white!
Populations tend to fluctuate. The Red squirrel
is now believed to be on the increase due to the
expanded area of afforestation. Generally they
are distributed all over the country except in
the midlands where the Grey squirrel has
replaced them and in the north-western areas
where the habitat is unsuitable.

Squirrels are easily recognised. The habit of
sitting on its haunches whilst gnawing at food
held in its front paws, with its extremely bushy
tail arched over its back, is unmistakable. There
is often confusion, however, when distinguishing

Red Squirrel, *Sciurus vulgaris.* Head
and body 220 mm in length. Well
distributed but more so in southern
half of country.

17

the Red squirrel from the Grey. The summer pelage is a rich, rufous brown with a darker, mid-dorsal band. The base of the belly is white, and in winter, its coat lacks much of the reddish hue, and may be ashy-grey by the next moult. There are prominent ear-tufts which are characteristic.

The great agility of the Red squirrel in trees is proverbial, appearing to be much more at home in such an environment. Food is generally taken to and eaten on a tree branch. The nest is known as a 'drey' and in a dense mass of twigs and bark lined with leaves, grass and shredded bark, placed in a branch fork. Gestation is about 46 days and the litter averages three, although from one to six are known. The young are born at an early stage of development with eyes and ears closed, and hair and teeth absent. Sexual maturity is said to be at between six and eleven months. The Red squirrel does not hibernate but merely goes through a period of restricted movement. Its food consists largely of conifer seeds, but a great variety of nuts, berries, fungi, bulbs and seeds, as well as sometimes the eggs and young of birds, are also taken. They can also strip the bark from trunks of forest conifers, so that it may well be a pest in plantations.

American Grey Squirrel, *Sciurus carolinensis*. Head and body average 260 mm, tail 215 mm in length. In eastern half of country but so far absent from the North and South.

American Grey Squirrel
Sciurus carolinensis

The Grey squirrel was introduced into Ireland at Castleforbes, Co. Longford in 1911, and has since spread into surrounding counties and into much of the midlands and over towards the east. It is generally considered to be a serious forestry pest, although it has many engaging ways. It differs from the native Red squirrel in that the winter pelage is grey with a yellowish-brown, mid-dorsal stripe, and it is a little longer than the summer coat, with white behind the ears. There is a rufous streak along the flanks which is the confusing feature — the inexpert believing it to be a Red squirrel. The tail has a white fringe in winter but this is not so marked in summer. Its preferred habitat is broadleaved or mixed woodland; indeed, the broadleaved element appears to be essential for feeding. The Grey squirrel possesses prodigious climbing ability. It is able to make 4m leaps, climb vertical, rough-cast walls and traverse horizontal wires. The 'drey', is made of twigs with attached leaves, lined with dry grass and leaves and it has a side entrance. It is located in a large fork, often at a considerable height from the ground, but a drey may be found in a hollow tree. There are 2 litters annually, gestation being about 44 days, whilst lactation is from seven to ten weeks. The average litter is three, although from one to seven have been recorded. Life-span is about five or six years.

Bank Vole *Clethrionomys glareolus*

Most voles are small, mouse-sized rodents related to the lemmings. The snout is blunt and the tail short, whereas the coat is long and dense. They also differ from other rodents in that the three molar teeth grow continuously just as do the incisors. Sharp ridges of hard enamel are subsequently produced, allowing the vole to cut up tough herbage. However, with the Bank vole, this condition lasts only a few months, when the molars produce 'roots' and grow no more. It then feeds on soft herbage and berries only.

It was thought until 1964 that, although there were about 100 species of vole found in the northern hemisphere with some in great abundance, there were none in Ireland, but the Bank vole is now known from an area in the

left: Bank Vole, *Clethrionomys glareolus*. Length of head and body about 100 mm and tail from 40-60 mm. Only recently discovered in Ireland and mostly bounded in the North by the Shannon and in the South by the River Bandon.

right: Wood Mouse, *Apodemus sylvaticus*. Head and body 85 mm in length. Generally distributed wherever woodland and shrubby areas occur but not on high ground.

below: House Mouse, *Mus musculus*. Length of head and body (male) average 80 mm, female a little less and tail in both sexes very slightly less than length of head and body. Found only in association with Man and his crops. Where Man is most abundant so is the House Mouse.

south-west (south of the Shannon and north of the river Bandon), but as yet this does not include high ground. The Bank vole appears to have been introduced about 150 years ago, although its reproductive rate is such that it could have arrived by some means or other only twenty years ago. In addition to the characteristics described, the ears and tail are furred. The back is chestnut with the belly silvery to creamy grey. The tail is only about half the length of the head and body. A pale, sandy-coloured variety is said to be not uncommon in Britain, but it is not known whether it occurs in Ireland.

The Bank vole's preferred habitat is broad-leaved woodland and scrub areas, where there is a degree of cover and where there is sufficient debris and herbs into which it can tunnel. It will,

however, emerge into the open from hedges, banks and stone walls. It is an agile runner and climber, and its voice consists of chattering and squeaking. The breeding season is generally from mid-April to September, but if the autumn is mild may extend to December. Gestation is 17 to 18 days with the number of embryos averaging 4.4 (in England), whilst the number of litters in the season is probably four or five. The young, which weigh only about 2gm. when born, are weaned at two-and-a-half weeks and are sexually mature at four or five weeks. The majority of individuals live only two seasons, including that of birth. The Bank vole feeds on soft vegetable matter, including grass and other green plants, but insects comprise about one-third of their total diet.

Wood Mouse *Apodemus sylvaticus*

Otherwise known as the Long-tailed Field mouse, this species is said to be almost certainly Ireland's commonest mammal. It is mainly nocturnal and, although widely distributed, seldom seen by ordinary folks, unless traps are set for suspected House mice in the larder or the apple store. At such time the Wood mouse may fall victim since it frequently enters buildings in autumn and winter (more especially when House mice are absent) and will consume a wide variety of human foodstuffs. It usually occurs in woodland and shrubby areas as well as in hedges, fields and gardens wherever there is coarse grass or some litter through which runways are constructed and also into the ground immediately beneath. The Wood mouse, rather larger than the House mouse, has yellowish-brown fur on the back and is white underneath and a yellowish-brown mark is normally present on the chest. It is more yellowish along the flanks. The eyes are large and conspicuous, being both round and prominently bulbous, and the ears are round and prominent also. The tail is only sparsely hairy and dark-brown above, whitish beneath and about equal in length to the head and body together. It makes a high-pitched squeak which is not often heard.

Breeding commences in spring and declines in October although it has been known to continue throughout the winter in some seasons. The nest is constructed below ground and consists of shredded grass. Above ground 'feeding platforms', sometimes well above ground, in old birds' nests, should not be confused with nests, although the latter are used for non-breeding purposes. Gestation is from 25 to 26 days, and the litter numbers about five or six. The number of litters per season is not known. Weaning takes place at about 21 days but the young first leave the nest at 15 or 16 days. Young born late in the season do not become sexually mature until the following year.

A wide range of food, both vegetable and animal, is consumed. Chief among the items is grain, seedling buds, fruit, nuts and snails. Adult and larval insects are commonly taken and from time to time winter wheat is grazed down to ground level and this makes the Wood mouse unpopular with farmers. A strange feature of this species is its habit of shedding the tail skin if held by the tail, except for the base. It will, indeed, shed the skinless tail later!

House Mouse *Mus musculus*

The House-mouse is now found world-wide. It has followed man wherever he has ventured and now lives in close association with him almost everywhere, not only in his buildings but also with his crops, although in adverse conditions it will often seek shelter indoors. This association with man is thought to have had its origins about 8,000 years ago when cereals were first cultivated. One would think that because it is familiar to almost everyone, that the House mouse would be readily identified by the householder, but, since the Pygmy shrew, the Wood mouse and, indeed, the Bank vole also enter buildings, mistakes in identification are often made.

The House mouse has a grey to grey-brown back whilst below it is grey to silvery-grey. Neither the eyes nor the ears are conspicuous and the sharp muzzle and long tail easily separate it from the Bank vole. It gives off a distinct odour easily recognizable even in moderate infestations, like that of the chemical acetamide.

Breeding takes place practically throughout the year, except when living out-of-doors. Nests are made of grass, paper, textiles of various sorts and a host of other materials which are capable

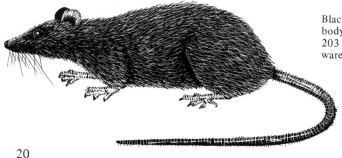

Black Rat, *Rattus rattus*. Male, length of head and body 190 mm and tail 220 mm, female 175 mm and 203 mm respectively. Found in upper stories of warehouses in ports but not on West Coast.

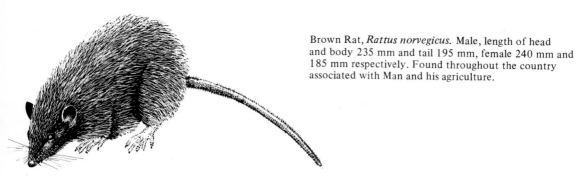

Brown Rat, *Rattus norvegicus*. Male, length of head and body 235 mm and tail 195 mm, female 240 mm and 185 mm respectively. Found throughout the country associated with Man and his agriculture.

of being shredded. Gestation is 19 to 20 days and the young are weaned at 18 days. Females become sexually mature at a weight of about 7.5gm., and males at about 10 gm. With abundant food and in warm conditions, five or six litters are produced annually, with five or six young per litter. The population does not increase indefinitely as, after a time, the males become territorial and prevent breeding by younger or less aggressive members of the group.

A number of colour varieties of the House mouse are known from Ireland. Chief among these is the sandy-coloured variety of North Bull Island in Dublin Bay. This is an area of sand-dune and saltmarsh and it is thought that the sand-coloured backs would give the mice more protection from bird predators such as the Kestrel.

Black Rat *Rattus rattus*

Known in Ireland, at least since 1185 when Giraldus Cambrensis mentioned it, the Black rat appears to have originated in South-east Asia. Although in Europe the predominant rat is the Brown rat, it is the Black rat which is the greatest pest on a world basis. The Black rat is much smaller than the Brown and possesses a noticeably more slender build. The tail is relatively longer and the ears larger, and the coat is not nearly so harsh as that of the Brown rat. Its feet are also more prehensile, making it a good climber, and indeed, whilst the Brown rat is usually on or under the ground, the Black is more at home in the upper stories of buildings. Thus it is often called the Roof rat.

As in the Brown, breeding can be continuous in the most favourable circumstances, but when

less so there are breeding peaks in summer and autumn. Gestation is 21 days with the number of embryos varying with the weight of the mother, who is sexually mature at about 90gm. which is from three to four months; at this weight about seven embryos are present. Between three and five litters are born annually.[2]

The Brown rat will generally drive away or kill the Black where their populations mix. Both species of rat are important carriers of disease. It was the Black rat which brought Bubonic Plague to northern Europe. This dreadful human scourge is caused by the bacterium *Yersinia pestis,* and it is carried by a number of species of rodent fleas. The biting of the fleas or the scratching of the flea-bite where infected faeces have been deposited cause the infection. The principal focus of the disease today is tropical Asia. Weil's Disease is caused by infection by Leptospirae. A number of animal species, including rats, can become infected and pass on the infection through their urine and faeces. Persons working in rat-infested areas (often sewer workers) are, therefore, at greater risk. Salmonellosis is a group of infectious diseases caused by species of the bacterial genus *Salmonella.* These are enteric in nature and human infection may be brought about by contact with rodents or with water and food contaminated by them.

Brown Rat *Rattus norvegicus*

Perhaps of all mammals, the Brown rat is loved least. Indeed, it is looked upon by most people, if not with horror, at least with disgust. It is a commensal of man, being almost wholly dependent on him for shelter and food. The

21

Brown rat is usually found in buildings, especially warehouses and factory premises where there are areas of little disturbance. In addition, it is often to be found in considerable numbers around farm buildings, on rubbish tips and in sewers. In summer, however, many take up residence in hedges and banks, making tunnels which are sometimes part of a complicated network. It is generally nocturnal. Being much larger than the House mouse, it can only be confused with the Black rat, *Rattus rattus*, but the Brown rat is about two and a half times heavier than the Black rat, reaching more than 500gm. on occasion. It is predominantly brown, although paler underneath, and the fur is coarse. Sturdy in appearance, it has a thick, scaly tail which is shorter than the head and body.

In a constant environment with a plentiful supply of food, breeding is continuous throughout the year with about 30% of the females pregnant at any one time. The bulky nests are made with any suitable material at hand. Gestation takes about 24 days and the number of embryos varies greatly according to the weight of the mother. At a weight of 160gm. there are about six embryos and at 500gm. there are about eleven. Between three and five litters are produced annually, and the young leave the nest after about three weeks. The females are sexually mature at approximately 150gm.

The range of foodstuffs eaten is very wide, being from almost 100% cereal wherever it is abundant, to a predominantly meat diet in places such as frozen meat stores. The Brown rat will kill and eat House mice by pulling the skin inside out. There is a very high mortality of young in the nest, and just on leaving the nest. When population is at saturation point the annual adult mortality is from 91 to 97 per cent. It was in 1722 that the Brown rat was first recorded as infesting the Dublin district, since then it has largely replaced the Black rat.

Carnivores *Carnivora*

This order of mammals primarily consists of flesh-eaters, although a number of species have secondarily-developed omnivorous eating habits or, in some cases, are vegetarian. The main anatomical features concern the teeth which have become highly specialized. The incisor teeth are generally reduced in size and are used for holding the prey. The canines are large and dagger-like and are used for killing the prey, whilst the cheek-teeth are usually reduced in size and number but have high, sharp cusps which are used for shearing the flesh and are known as carnassials. Claws may be retractile, as in cats, or non-retractile or semitractile. Five toes occur on each foot in the family *Mustelidae,* but only four in the dogs and cats.

Fox *Vulpes vulpes*

Although said to be rare a hundred years ago, the fox today has a wide distribution throughout the Irish countryside, and as a raider of dustbins and rubbish dumps it is a visitor to the fringes of built-up areas. Its appearance is well-known: the pointed muzzle (mask); the sharp, upstanding ears; the reddish-brown coat with the long, white-tipped, bushy tail (brush) make a familiar sight to the countryman who is abroad early in the morning. The sexes are similar although the vixen is said to be somewhat smaller and her face narrower than that of the dog. The five claws on the forefeet and the four on the hind are not retractile and may be seen in the tracks together with the marks of hairs which grow between the pads. The head and back of an average adult measures about 65cm whilst the tail is another 40cm. The dog weighs about 6.8kg and the vixen about 5.5kg. Foxes are usually solitary except when breeding, and hunt around their 2,000 hectare (one square mile) or so of home territory. They proceed at a walk or a trot but will also gallop. The latter is usually seen when they cross a road. When hunted, they are able to run for considerable distances, using a number of behavioural devices, such as seeking a fresh fox, when confusing the hounds.

Breeding takes place from early January to February, and although the vixen is receptive for about three weeks, for only three days is she capable of being fertilized. Gestation is 51 to 52 days and the average number of cubs in a litter is 4.7 with only one litter per season. The litter is produced in an 'earth' which may be an enlarged rabbit burrow or an old badger set. Several are cleaned out before selection is made,

right: Fox, *Vulpes vulpes.* Length of head and body 650 mm, tail 400 mm in the male with the female slightly smaller. Widely distributed throughout the country and visits outer urban areas.

left: Irish Stoat, *Mustela erminea hibernica.* Length of head and body (male) about 260 mm, tail 94 mm, female 222 mm and 69 mm respectively. Common throughout Ireland with very few areas where it has not been recorded.

below: Badger, *Meles meles.* Average length of head and body 815 mm, tail 100 mm. Common and widespread wherever suitable soil depth occurs for excavation of a 'set'.

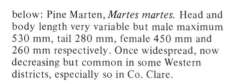

below: Pine Marten, *Martes martes.* Head and body length very variable but male maximum 530 mm, tail 280 mm, female 450 mm and 260 mm respectively. Once widespread, now decreasing but common in some Western districts, especially so in Co. Clare.

and the vixen may transfer her cubs from one to another if there is any disturbance. After about three and a half weeks the cubs emerge from the earth and are weaned at eight to ten weeks. They are sexually mature in December following birth (about nine months).

The food of the fox is largely conditioned by availability. When rabbits are abundant they constitute a large component of the diet. However, insects are taken in summer as well as vegetable matter in the form of fruit and berries which are eaten from spring to autumn. The fox is unpopular for taking gamebirds and domestic poultry — usually in spring and summer — but rodents are also consumed, mostly in the autumn and winter. The Brown rat makes a larger contribution to the foxes' diet in Ireland than in England.

Pine Marten *Martes martes*
At one time, this attractive carnivore was to be found throughout Ireland but was then believed to occur only in about ten counties with the total population consisting of not much more than about two hundred animals. This decline was put down to it being hunted for its valuable fur.[3] The Pine marten is mainly a woodland animal inhabiting mixed and coniferous forest, although it may also be found on rocky or rough scrubby slopes, crags and screes. It is extremely agile and is able to catch a squirrel in a tree.

The Pine marten's general form is like that of a very large stoat in that it has an elongated body with short legs, but the tail is long and bushy. Its fur is rich brown to almost black with a cream patch on the throat extending to the chest. However, when freshly moulted this may be tinged with orange. The ears are conspicuous and their margins are pale. The length of the head and back of the male may reach 530mm with the tail measuring 280mm. The female's corresponding figures are 450mm and 260mm. The largest males may weigh up to 1.5kg.

Its food consists principally of rodents and small birds. Of the rodents, voles are probably an important component of the diet where they occur, whilst wrens and tree-creepers are common amongst the birds taken by the marten. In Sweden, squirrels form an important part of

the winter diet. Berries are also eaten when available. The Pine marten has a den, usually in rocky crevices, although magpie nests are sometimes used. Copulation is a rather tempestuous affair lasting about an hour with the male dragging his mate around by the scruff of her neck with much purring and growling. Mating takes place around July and August but implantation is delayed until mid-January. The young, probably about three, are born in late March to April. They are weaned at about six to seven weeks and emerge from the den at about two months old.

Irish Stoat *Mustela ermines hibernica*
The stoat in Ireland is often referred to as the 'Weasel', but the latter name is correctly assigned to the much smaller *Mustela nivalis* which does not occur in Ireland. The stoat, although generally nocturnal, often hunts by day, so is likely to be encountered, not only by the countryman, but also by the motorist away from towns for they are often seen to bound across the road. They occur throughout Ireland.

The elongate body, short legs and reddish-brown fur make the stoat unmistakable in Ireland as no other species is similar. The Irish stoat, the subspecies *hibernica*, differs from the British subspecies *stabilis*, in having no white margin to the ears, no white on the upper lip and having considerably less creamy-white on the belly. It is also slightly smaller than the British subspecies. The male has a head and body length of between 24.0 and 28.3cm, and the female between 18.4 and 26.0cm. The tail, which has a black tuft of hair at the tip is from 7.2cm to 11.7cm in the male, and 5.7 to 8.1cm in the female. In the past it was thought that the Irish stoats had five or six pairs of teats and the British stoats four pairs. In fact, both subspecies may have from four to six pairs.

Whilst the males are sexually active from March through to November, fertilization can only take place in late spring to summer and implantation is delayed until the spring following, the delay being about 280 days. The embryos then take a further 21-28 days to complete their development. There is one litter only each year. Females are fertilized by older males during their first summer, but males are not sexually mature until the year after birth.

The stoat is a fierce carnivore, attacking and devouring mammals much larger than itself, and rabbits and rats are no match for it. Mice and voles are taken as well as birds and their eggs, shrews, reptiles and fish. Berries, however, may also be taken in quantity in the autumn. There is a wealth of anecdote concerning the stoat — its inquisitiveness, its 'herding' into packs, its aggression, the sounds it makes, its curious playing, together with other aspects of its biology and behaviour are all of interest, and the reader is referred to J. S. Fairley's *An Irish Beast Book* for a lively account of the stoat's activities.

American Mink *Mustela vison*

North American mink were introduced into Ireland for fur-farming during the 1950s and during the 1960s many escaped. It is possible that they now thrive in a feral condition in Ireland. Seldom found far from water, the mink's food consists of small mammals, frogs and insects as well as some vegetable matter such as buds and berries. The fur is dark brown except for a small white spot on the lower lip and the chin, although recently-escaped mink may be from white to near-black. Records of sightings up to 1973 show it to be distributed in the north, the east and in Kerry in the south-west, but it has spread further.

Badger *Meles meles*

In spite of much persecution, the badger occurs in every county throughout Ireland wherever suitable habitats occur. The fringes of woodland with a soil depth sufficient for the excavation of the 'set', the burrow where breeding and sleeping takes place, are the badgers' favourite sites, and sets are rarely found in marshy areas. Unfortunately, much of its harrassment is due to the ill-founded reputation it has of preying on domestic poultry and game birds. In fact the larger part of the badger's diet consists of earthworms, injurious insects such as cockchafer beetles, and rodents. The badger, therefore, is actually beneficial! It may, however, make a

nuisance of itself by rolling in corn in order to get at the ears which make a tasty variation to its usual food.

In general appearance the badger is squat with short legs and a relatively massive body. The fur is greyish although individual hairs are dark in the centre and paler at the base and tip. The wedge-shaped head is white, with two black bands extending from the muzzle, through the small eyes and white-tipped ears to the neck. The short, powerful legs have five unretractile claws on each foot with those on the forefeet long and well adapted for digging. The average weight of an adult male badger is 12.3kg, and the female 10.9kg.

The paths used by badgers may be identified by the hairs caught in the lowest strands of barbed-wire where the animals have pushed underneath. Although occasionally found abroad in daylight, it is mostly nocturnal, spending the day in its set which is a complex, underground tunnel system. The set is often located by the presence of old bedding — bracken, dry grass and leaves. This is dragged from the set before being replaced with fresh material. In diet it is omnivorous, eating what is generally available, but earthworms make up the largest single coponent. Young rabbits, rodents, moles, hedgehogs, frogs, slugs, snails, wasp grubs, and beetles are also eaten, and wasp nests are frequently dug out completely! Sometimes eggs of ground-nesting birds are taken, and occasionally those of poultry. Carrion, such as dead young rooks fallen from the nest, is eaten and a wide variety of vegetable material. Fruits, acorns, bulbs, windfall apples, plums and blackberries, oats and wheat are also devoured.

For many years the breeding habits of the badger were an enigma. Mating takes place throughout spring, summer and autumn, although fertilization usually takes place from February to May only. There is a long period of delay before implantation which may be as much as nine months and is usually in December. During the implantation delay the sow will receive the boar. The one to five young are born from January to May, but sometimes larger numbers have been recorded due to two litters being produced in one set. The cubs are about eight weeks old before they first emerge, and they commence to wean at twelve weeks, when the sow feeds them by regurgitating semi-digested food. The cubs stay with the sow until autumn and often overwinter with her.

Otter *Lutra lutra*

Until recently considered a common animal around lakes and rivers (and also around the coast), the otter is now possibly diminishing in numbers, as has occurred in Europe generally and particularly in England. It is a most attractive animal because of its natural grace and beauty of lithesome movements when in water — a habitat for which it is highly adapted. Unfortunately it does not enjoy a good reputation with fishermen, since it competes with them for the fish. Its fur is valuable and it is often persecuted for this too.

The otter has an elongated body and short legs. The head has a flat appearance and the tail (the rudder), is thick at the base and tapers to a point. The fur is dense and brownish and it has a characteristic spiky appearance when the animal emerges from the water. This is because the guard hairs stick together when wet. The feet are webbed, but the fifth toe some-times leaves a mark in the footprint (known as the 'seal'). The total length of the male is about 118cm whilst the female is about 104cm. The male weighs about 10.3kg and the female about 7.4kg.

Little is known of the breeding habits of the otter but delayed implantation occurs in related species, so that the picture of the breeding pattern is not clear. The nest of the otter is known as the 'holt' and is usually made in a drain or in a hollow tree and lined with reeds, dried grass and moss. It may be at some distance from the main stream. There are many records of breeding spread throughout the year but gestation appears to be about 62 days. The young are born blind and remain so for 35 days, staying in the holt for about two months, and probably remain with the bitch until she mates again.

Otters are usually nocturnal and are always on the move, seldom stopping in any one place for long, unless young are in the holt. Generally, there are too few otters to cause serious concern to fishery interests, except in the case of fish hatcheries at spawning time, when considerable damage may result. However, eels are a favourite food item, as are crayfish. In addition to frogs, small birds, fish, rodents and insects are also taken, together with newts, slugs, earthworms, tadpoles and such small items as freshwater shrimps. Around the coast otters take crabs and Sea-urchins, as well as many other marine animals whose remains are often found on the shore. Defaecation by the otter is known as 'sprainting'. This is often carried out on the top of a rock or a mound, and sometimes sand or earth is scraped into a mound as a preliminary. The 'spraints' are black with a characteristic odour which is not unpleasant.

'Odd-Toed' Ungulates *Perissodactyla*

In these mammals, the weight of the body is carried either on one digit of each foot — such as in the horse and its allies — or on three digits, as in the rhinoceroses or the tapirs. In Ireland the sole representative of this order is the horse.

Otter, *Lutra lutra*. Length, inclusive of tail, about 118 cm (male), and female about 104 cm. Well-distributed throughout the country but now decreasing in numbers.

Connemara Pony *Equus caballus*

The origins of the breed of semi-wild pony found in Connemara and some adjacent districts are lost in antiquity. These ponies lived for hundreds of years in a feral state and entirely by their own efforts in wild and difficult country. Some authorities consider that the Connemara pony came ashore from wrecked Spanish galleons in 1588, but others are convinced that the breed had been established hundreds of years previously, and that it is a Celtic-type animal having affinities with the ponies of Norway, Shetland and the Scottish Highlands.

It is extremely hardy in constitution and stands at a height of 13 to 14 hands (1.39 – 1.45m or 52-56 inches). The body is compact and the legs are short – covering a lot of ground. The typical colour of the breed was at one time dun, but the predominant colour is now grey, with dun-coloured animals being very scarce. Blacks are a little more numerous than are browns and bays. The native-bred Connemara is a wiry, versatile pony, able to thrive on poor ground.

Even-Toed Ungulates *Artiodactyla*

This is one of the most important orders of the world's mammals and consists of large, vegetarian browsers and grazers. They are unaggressive and escape from would-be predators by swiftness of foot. The chief characteristic is the possession of equally developed third and fourth digits, with the limb axis passing between them. Horns and antlers have also been developed, but upper incisor teeth are either much reduced or absent. In the case of cattle, herbage is plucked by the tongue, whilst deer pull the herbage by the lower incisors pressing against a pad in the upper jaw. Those species present in Ireland are as follows:

Red Deer	*Cervus elephas*
Fallow Deer	*Dama dama*
Sika Deer	*Cervus nippon*
Feral Goat	*Capra hircus*

Horns carried by cattle, goats and some breeds of sheep, have a bony core and are not shed annually but grow from year to year. The antlers, carried only by male deer, are shed annually. New antlers grow and increase in size each year, until a maximum size is attained, when they start to 'go back' or degenerate.

Fallow Deer *Dama dama*

Fallow deer are found locally in many parts of Ireland but those living 'wild' are descendants of animals which have escaped (or have been allowed to escape) from parks. It is not a native species but was probably introduced by the Normans. The male (buck) is instantly recognizable by its flattened (or palmate) antlers, from which arise a number of small spikes known as 'spellers'. Standing at about 95cm, it is much smaller than the Red deer stag, and the antlers reach a length from 65 to 75cm. A clean carcase weight is from about 64 to 110kg. Its coat varies widely in its colouration but is generally a dark fawn spotted on the flanks with white. There is a line of black hairs down the middle of the back extending to the tail. Animals which are almost black and with the spotting almost invisible, are often met with, and a herd of white Fallow occurs at Grangecon, Co. Wicklow.

Antlers are grown in the second year, the first being simple spikes. Such an animal is known as a 'pricket'. The antlers reach their maximum development in the sixth year, being cast each May and by August having regrown with a covering of 'velvet'. This is a highly sensitive tissue abundantly furnished with blood vessels. When the antlers are grown fully, this tissue of velvet dies and the buck rubs it off against tree branches and shrubs. The antlers are usually clear of velvet by the end of August. The does or females do not bear antlers.

Mating time, known as the 'rut', commences in the middle of October when the senior buck or 'master' buck takes up his mating territory, defending it against intrusion by other bucks with great vigour. The master buck frays young trees, makes 'scrapes' by them and urinates in them, ultimately transferring this odour to his antlers, together with scent from the post-orbital glands which are situated near the eyes. This scent is then dispersed and picked up by the does who then congregate in the rutting area. During the rut the master buck produces a deep-throated grunt. The rut lasts about one month. Does may become pregnant when two years old and the single fawn (rarely are there two) is born in May or June, although November has been recorded. Gestation is eight months.

Sika Deer *Cervus nippon*

The Japanese sub-species of this far-ranging deer was introduced into parks in the British isles in the mid-nineteenth century.[4] Many subsequently escaped but were able to look after themselves in surrounding woodland very well indeed. In Ireland, Sika occur in the feral state most abundantly around Killarney, Kilgarvin and Killorglin in Co. Kerry, and cause much damage to regenerating forest. In Co. Cork they are found in Glengariff and they have become established in Fermanagh, Tyrone, Wicklow and Dublin.

Only very slightly smaller than Fallow (this would not be noticeable in the field) the Sika has buff-brown flanks with faint spotting in summer. It is dark brown in winter. The head is greyer in colour and the tail is white, making it easier to distinguish from the black-tailed Fallow deer. The antlers resemble those of the Red deer, but they are simpler and much slighter and are cast early in April. The new antlers are in velvet during the whole of the summer, of a rich-red colour with black tips. They are free from velvet by the end of September. The length of the antlers varies from 40 to 63 cm.

The rut commences at the end of September and continues through to the end of November. During this time the stag calls with an easily recognizable whistle which ends in a grunt. Each stag herds five or six hinds, and the calves are born late in May. An unfortunate consequence of Sika mixing with a population of Red deer is the hybridization which takes place. Hybridization between Red and Sika deer has taken place in Wicklow and Dublin, and in Donegal, Tyrone and Fermanagh. In Kerry Red Deer tend to keep to the mountains and Sika to the forest.

Red Deer *Cervus elephas*

Red deer have a good claim to be considered indigenous Irish animals, and these truly wild

Fallow Deer, *Dama dama.* Height at shoulder of Buck, 90-95 cm. Doe slightly smaller. Antlers up to 78 cm in length. There are 8 Park herds but feral animals common in some districts but absent from extreme North, South and South-east.

Sika Deer, *Cervus nippon.* Height at shoulder of Stag, 82-90 cm. Hind, 68-75 cm. Antlers up to 63 cm in length. Occurs not uncommonly in woodland areas of Counties Kerry, Cork, Fermanagh, Tyrone, Wicklow and Dublin.

above: Irish Giant Deer. Antlers of Buck. Total span up to 4 m.

left: Red Deer, *Cervus elephas.* Height at shoulder (Stag) up to 140 cm, Hind about 105 cm. Antlers about 100 cm in length. Occurs locally in Co. Donegal (most numerous), Killarney and Wicklow Mountains.

right: Irish Giant Deer, *Megaloceros giganteus.* Height at shoulder 2.5 m. Skeletons found in areas which were open, including lake sides and marsh from some 12,000 years ago.

deer occur most numerously in Glenveagh, Co. Donegal and less so around Killarney and in the Wicklow mountains. In addition, they are found in a number of deer parks in many parts of the country. Larger than Fallow, the male (stag) stands about 105 to 140cm at the withers. Antlers from British specimens average about 100cm in length. The clean weight of a Scottish stag is approximately 95kg, but English woodland specimens may weigh up to 189kg. In summer the coat is dark red to brown and the inner thighs are whitish-cream. The tail is usually furnished with a black, central streak. However, some stags do not possess antlers (a hummel), and others bear no points above the brow-tine (a switch). In April the antlers are cast and by August new antlers are at maximum size, and are clear of velvet by the end of August. The females do not bear antlers.

The rut takes place during October and November when the stag wallows in muddy pools and 'roars' incessantly. The females (hinds) become pregnant in their third year and bear their first young (calf) in their fourth. Gestation is eight months, and the calving period is from the end of May to early in June. The single calf is suckled for eight to ten months and remains with the hind until its second autumn.

Irish Giant Deer *Megaloceros giganteus*

No account of the fauna of Ireland would be complete without some notes on the Giant Deer of Ireland, although it seems certain that man never caught sight of this magnificent animal. All the evidence appears to show that it became extinct about 10,000 years ago, and man arrived in Ireland approximately 2,000 years later. The Giant Deer flourished in Ireland for a relatively brief period — about a thousand years or so. It made its way from Continental Europe when there was a land connection, because so much water was locked up as ice, and they were able to walk across dryshod. This was towards the end of the last glaciation when a warm phase occurred, sandwiched between two colder ones. This gigantic deer had become abundant in open, grassy country which was only sparsely wooded, but when conditions changed and Arctic Tundra crept remorselessly southwards,

the deer was unable to adapt itself to new conditions of frozen bogland or, when the ice sheets receded, dense forests.

The Irish Giant Deer stood about 2.5m at the shoulder, but the extraordinary feature of the animal is not so much its body size as the disproportionately large antlers and their strange shape. They are the largest antlers ever known and estimates of their total span range up to 4m.

In shape they are fallow-like, having a broad, expanded palm. The brow-tines are small, expanded somewhat, and variable, and their closeness to the head gives the impression that they had no function. Between the latter and the palm is another tine, also variable in shape and size and often flattened. The palm is enormously expanded with five or six long, backwardly-directed spellers (in Fallow language). There is one important difference, however, between the antlers of the Fallow and those of the Giant Deer. In the Fallow, the beams of the antlers are directed upwards so that they are vertical when the head is held upright, but in the Giant Deer the beams turn outwards so that the antlers are horizontal when the head is upright. Much has been written concerning their function because they would be useless for defence against wolves and other predators since they point backwards instead of forwards. Once, it was generally accepted that they were for courtship only — the buck would wave them up and down in front of the antler-less doe. Now, it is held that these gigantic antlers determined social ranking within the herd and, indeed, made fighting unnecessary.

Wild Goat *Capra hircus*

Strictly speaking the Wild goats of Ireland are feral goats in that they have been derived from domestic animals which have either escaped or have been deliberately set free. However, this may have occurred in early times even though such animals revert to a wild type within a few generations. These animals are certainly wild in that they shun any association with man and usually eke out a very precarious existence in inhospitable, mountainous country difficult of human access. They are helped in this by a muscular adaptation of their hooves whereby

Distribution of the Wild Goat in Ireland.

they are able to grip with the pair of hoofed digits. When spread, the digits can then be pulled together and this enables the goat to hold its position on the slightest tuft of grass or imperfection on a rock face. This has helped the goats to become elusive. Until the Second World War these herds were scarcely molested, but that time saw many herds completely destroyed or very considerably reduced in numbers so that their existence is now open to question.

'Wildness' in the goat, apart from behaviour, is usually expressed in 'hair and horn'. Its coat becomes long and shaggy and the colour varies from creamy-white, black and brown; brown and white or black and white. The horns, which are never shed, continue to grow from year to year, having started growth almost parallel but subsequently diverging until the tips are almost pointing in opposite directions. The age of the animal can be calculated from the 'growth rings' on the horns.

NOTES
1. The weight of the Pygmy Shrew varies according to season and ranges from 2.5 to 5 gm, and an odd one may attain 6 gm.

2. The distribution of the black rat is confined chiefly to seaports (Derry, Belfast, Carlingford, Dublin, Cork, etc.) but there are isolated (unsubstantiated) reports from a number of other localities around the country, chiefly from the north-east.

3. At the moment, however, it is believed that Pine martens are much more numerous than is generally thought. New records are coming in and the distribution is wider than the ten counties. The possible reason for this may be afforestation.

4. Sika was introduced into Ireland by Viscount Powers about 1860 and released in Powerscourt Park. Later it was sent elsewhere in Ireland.

Goats clear land of much noxious plant life, such as bramble, briar, ivy, gorse and heather and will eat thistles, docks, stinging-nettles and many other weeds. When great numbers of goats were removed from the Burren in Co. Clare, such plants returned to areas where cattle-grazing had previously taken place and this reduced the number of stock able to be carried. On the other hand, goats will remove all seedling trees by their close cropping, thus making the regeneration of woodland virtually impossible. Indeed, this is the most important crime placed at the feet of the Wild goat.

Whitehead has listed the goat herds of Ireland, from which the following notes are taken:

Clare: (Burren)	Only a few remain.
Donegal:	A few frequent the Glenveagh district.
Dublin:	Not truly feral animals on headland opposite Howth and Baily lighthouse.
Galway:	Costelloe: About 100 on the lake shores between Leam and Costelloe, often swimming out to the Islands when disturbed.
	Joyce's Country: Small herds in mountains of Northern Connemara.
	Twelve Pins Mountains: A herd of about 50 near Kylemore Lake.
Kerry: (Skellig)	A few said to exist on the Skellig Rocks. About 150 goats established on Mt. Eagle and Mt. Brandon, and a herd of white goats reported on the Dingle Peninsula. A few on Torc and Mangerton Mountains.
Mayo:	The old, blue-grey herd now extinct. The present long-haired whites are of recent origin.
Sligo:	A few still remain.
Tipperary:	No recent information.
Waterford:	No recent information.
Wicklow:	Not truly feral on Bray Head. About 60 on cliffs overlooking Glendalough Lake.
Antrim:	A few on Fair Head and Garron Point, Glenarm and Rathlin Island.
Armagh:	About 20 on Camlough Mountain and a few, small herds in Slieve Gullion.
Down:	Perhaps now extinct in Mourne Mountains.
Fermanagh:	Small numbers on islands in Lough Erne and on cliffs in Lough Navan region.

2 Mammals of the Sea

Whales, Dolphins, Porpoises, Seals

WHALES, PORPOISES, DOLPHINS *Cetacea*

The large members of this order of mammals are known as whales, no matter to what group they belong. In fact, species larger than 6.5m (20ft.) are thus named, and the killer whale, *Orcinus orca,* for instance, is actually a dolphin!

The *Cetacea* includes the largest living animals, and all are highly adapted for a marine existence. They are fish-like in general shape and are born, feed and reproduce, suckle their young, sleep and die in the sea. Those found on beaches have met their death due to accidental displacement from their natural environment. Some species are more prone to do this than others. It does seem a great pity that we are unaware of these huge animals living their lives out or just passing through on their migration in Irish waters, until they are found dead. This is either accidental or when they have fallen prey to man when their exploitation was common. However, many species of whale, happily, are conserved by international agreement on account of the fact that their numbers have decreased to such a low level that their very continued existence remains in doubt.

The various anatomical modifications brought about by a complete aquatic life include the absence of a neck region; the transference of the nostrils from the tip of the snout to a position usually at the top of the head: the enclosing of the digits of the forelimb which,

with other modifications, has resulted in fin-like limbs as flippers, and the absence of any external evidence of hind-limbs but with the body terminating in a pair of horizontal, fleshy fin-like processes known as flukes. Other modifications include the openings for the ears being hardly discernible; the possession of a fleshy, dorsal fin unsupported by any bony structure and the mammary glands being concealed in slits. Hair is confined to just a few on the head and the jaw of adult Baleen whales, but otherwise it is found only on the immature toothed species. Body temperature is maintained by a thick layer of blubber. Propulsion is brought about by the vertical thrusts of the tail region and the flukes.

The compilation of a list of whales occurring in Irish waters is a difficult task. Whereas some dolphins and porpoises seem to enter harbours and estuaries quite commonly, most whales usually keep well out to sea. With the extension of the limits of territorial waters, it is reasonable that more should be known of their largest inhabitants. Unfortunately, in spite of their impressive size, little is known of them, and not only have many species become exceedingly rare due to their exploitation in whaling days, but even when sighted they are often difficult to identify. It is indeed tragic that we know what we do of the occurrence of whales in Irish and British waters from the stranding of dead

Classified list of Irish Cetacea
Whales, Dolphins, Porpoises

Suborder MYSTICETI
Family BALAENIDAE

North Atlantic or Biscayan Right Whale
    Balaena glacialis
Humpback Whale
    Megaptera novaeangliae
Common Rorqual or Fin Whale
    Balaenoptera physalus
Lesser Rorqual or Pike Whale
    Balaenoptera acutorostrata
Blue Whale or Sibbald's Whale
    Balaenoptera musculus
Sei Whale or Rudolphi's Rorqual
    Balaenoptera borealis

Suborder ODONTOCETI
Family PHYSETERIDAE

Sperm Whale or Cachalot
    Physeter catodon
Pigmy Sperm Whale or Lesser Cachalot
    Kogia brevicepts

Suborder ODONTOCETI
Family ZIPHIDAE

Bottle-nosed Whale
    Hyperoodon ampullatus
Cuvier's Whale Ziphius cavirostris
Sowerby's Whale Mesoplodon bidens
True's Beaked Whale Mesoplodon mirus

Family MONODONTIDAE

Common Porpoise Phocoena phocoena

Family DELPHINIDAE

Killer or Grampus Orcinus orca
False Killer Pseudorca crassidens
Pilot Whale or Blackfish or Ca'aing Whale
    Globicephala melaena
Risso's Dolphin Grampus griseus
Bottle-nosed Dolphin Tursiops truncatus
White-beaked Dolphin
    Lagenorhynchus albirostris
White-sided Dolphin
    Lagenorhynchus acutus
Common Dolphin Delphinus delphis

animals. Why whales should meet their end in this way, occasionally in considerable numbers, remains something of a mystery.

In 1913, a scheme of recording the stranding of whales and other cetaceans around the Irish and British coasts was initiated by the British Museum. With the help of the Department of Lands and Fisheries, and the National Museum of Ireland in Dublin and the Royal Scottish Museum in Edinburgh, a total of 1810 identifications have been made up until 1974. Perhaps one day whales, with continued protection, may become more numerous around our waters and we shall be able to promote studies of living specimens. Already the sounds made by them during migration in North American waters have been studied, so that perhaps we may look forward to similar studies being made on the Irish side of the Atlantic.

Biscayan Whale *Balaena glacialis*
Also known as the North Atlantic Right whale, this species was at one time abundant from the Bay of Biscay to Newfoundland and throughout the southern areas of the Arctic seas. Now, alas, it is rare. It can only be hoped that sufficient numbers exist for it to re-establish itself throughout its old haunts. Eighteen specimens were taken between 1908 and 1910 by the Inishkea and the Blacksod Whaling Companies. The head of this whale takes up about one quarter of the whole length and the blow-holes are a pair of diverging slits. At the tip of the upper jaw there is a horn-like excresence known as the bonnet and it is generally well-accommodated with whale-lice, barnacles and worms. Although the body colour is black, occasionally white irregular patches are to be seen on the belly. It measures up to 18.5m in length with the black baleen blades up to 2.7m in length. In winter it appears to make its way to North Africa and in summer travels to Arctic waters.

Humpback Whale *Megaptera novaeangliae*
This 15m long whale used to be part of whaling catches in the early years of this century. The most characteristic feature is that of the flippers

left: Biscayan Whale,
Balaena glacialis. Up to
18.5 m in length. Rare. In
territorial waters when
on migration.

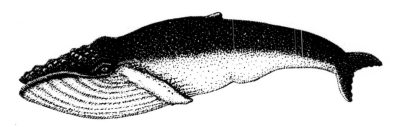

right: Humpback Whale,
Megaptera novaeanglia. 15 m
in length. In territorial waters
numbers unknown.

left: Common Rorqual,
Balaenoptera physalus. Reaches
24.5 m in length, the female
is the larger. Migrates in territorial
waters. Numbers unknown.

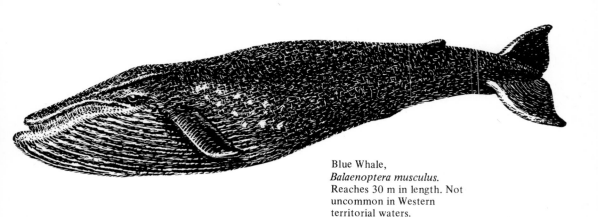

Blue Whale,
Balaenoptera musculus.
Reaches 30 m in length. Not
uncommon in Western
territorial waters.

35

— from one-quarter to one-third of its body length, and irregularly knobbed. The jaws are similarly humped and the dorsal fin is situated on a low hump at the beginning of the hinder third of its body. The flukes are notched in the middle. The baleen is black and about 1m in length. This whale shows great variation in its behaviour at the surface — rolling, side-swimming, breaching clear of the water and making a series of shallow dives followed by a deep dive (sounding).

Common Rorqual *Balaenoptera physalus*
Also known as the Fin whale, this species migrates northwards along the coast in late winter or early spring, returning southwards in autumn. There are four species of Rorqual *(Balaenoptera)* found in the waters off Ireland, and all have about 85-90 grooves in the throat and chest. The Common Rorqual is up to 24.5m in length with the female larger than the male. It congregates in 'schools' which may number up to 200 individuals. The dorsal fin is high, triangular and has a concave margin. Body colour is grey on top and white underneath, but colouration of the head is asymmetrical. The baleen plates are about 1m in length at the most. All Rorquals 'blow' vertically and the flukes are not usually visible when diving.

Lesser Rorqual
Balaenoptera acutorostrata
Otherwise known as the Pike whale or Minke whale, this is the commonest of the Rorquals likely to be seen off the Irish coast. It is also the smallest of them, the largest being 9m in length. It migrates as the Common Rorqual. The flippers are about one-eighth of its body length, and there are about 50 throat grooves. The general colour is dark, greyish-brown to black on the upper parts and the underparts white. A white patch on the flippers is distinctive.

Sei Whale *Balaenoptera borealis*
Known also as Rudolphi's Rorqual, it is not very abundant but may be encountered in Irish waters in June.[1] Its flippers are about one-eleventh of the total body length which is usually about 15m. There are forty throat grooves and some hairs are present on the upper and lower jaws and the chin. The general colour of the Sei whale is bluish-grey, with a lighter grey underneath. The 'blow' is inconspicuous, and this species exposes itself very little when surfacing.

Blue Whale *Balaenoptera musculus*
This is the largest living mammal and is otherwise known as Sibbald's Rorqual. Said to be not uncommon on the western seaboard of Britain and Ireland,[2] it reaches up to 30m in length and the weight of this huge animal is in the region of twenty tonnes. There are between seventy and 120 throat grooves and a few hairs are present on the chin and rostrum. The ground colour is dark, bluish-grey and although there are no large unpigmented areas, there are numerous white splashes which may aggregate here and there. However, pigmentation is absent from the under-surface and from the tip of the upper surface of the flippers. The 'blow' is high. In spite of its enormous size, the Blue whale is easily scared by strange sounds. On diving, it does not usually show the tail flukes.

TOOTHED WHALES
Suborder Odontoceti

Sperm Whale *Physeter catodon*
Also known as the Cachalot, this is the largest of the toothed whales, and although it is usually regarded as occurring in tropical and subtropical seas, it is known to occur around the coasts of the British Isles as 11 animals have been stranded since 1913.[3]

The head is disproportionately large, being about one-third of the total body length, and the upper jaw is square and transverse. The lower jaw does not extend as far forward as the extreme front of the upper, being mostly very narrow with the teeth conspicuous whereas they are concealed in the upper jaw. The flippers are rounded, but the dorsal fin and the succeeding four or five humps are not well-defined. There is also an irregular corrugation of the ventral surface of the body region, the colour of the latter being generally dark grey to black with the belly whitish with much variation in its extent. In the lower jaw there are from twenty to thirty teeth, and the teeth in the upper jaw are not functionable.

The length of the male is up to 18m, but the female is only half this size. The weight of a fully-grown male is about 50 tons! The 'blow' is directed forwards, and when surfacing the back fin is observed first and then the head comes out of the water. When sounding, the body arches and the flukes break clear of the water before the body submerges vertically.

Bottle-nosed Whales *Ziphidae*
These whales are medium in size but they possess a more or less conspicuous snout. There is no notch in the hind margin of the tail-flukes and there is only a single blow-hole.

Bottle-nosed Whale
Hyperoodon ampullatus
This fairly common, 9m or so whale, is thought to migrate around the western coasts and through the Irish Sea, and in its southwards route, in late autumn and winter, it comes to grief in strandings. The snout is about 15cm long and the forehead bulges increasingly with age. The colour of the body is dark grey to black and underneath there are variable amounts of light grey to white. There is a single pair of teeth at the very tip of the lower jaw in the males. In the female, teeth are concealed but vestigial teeth are embedded in gums. Schools of about a dozen are usually encountered and they are presumed to feed mainly on cuttlefish and herring.

Cuvier's Whale *Ziphius cavirostris*
This is also known as the Goose-beaked whale, and practically nothing is known about it, although at least eleven have been stranded on Irish south and west coasts from 1913 until 1974. Its length appears to be about 8m and the colour is said to be bluish-black with long scars on the skin. Only one pair of teeth is exposed at the tip of the lower jaw and only in the male,

but usually vestigial teeth are present.

Sowerby's Whale *Mesoplodon bidens*
This species is comparatively rare although there have been four strandings around the Irish coast — two since 1913. Almost nothing is known about it, except that it is mostly black with some white underneath and the skin is usually heavily scarred. Only a single pair of teeth project from about halfway between the tip of the lower jaw and the angle of gape. In the male, these teeth are flattened and triangular but in the female they are concealed. It measures up to about 5m in length.

True's Beaked Whale *Mesoplodon mirus*
Of all the whales of the world, this is among the rarest. Only fourteen are known from the whole world but four of these were stranded on Irish shores. It is about 5m in length and the single pair of flattened teeth are located at the extreme end of the lower jaw and point obliquely forwards. In the female the teeth are concealed.

Common Porpoise *Phocoena phocoena*
Also known as the Atlantic Harbour porpoise, this is probably the most abundant cetacean occurring in Irish waters. It reaches a length of only about 1.8m, is stout and beakless and has a receding forehead. The back is black and the belly white — the two merging into lateral, grey areas. The mouth is black, as are the flippers and flukes, and a dark streak joins the flippers with the mouth. There are between 22 and 27 spade-shaped teeth on each side of both upper and lower jaws. The Common porpoise is not a fast swimmer and never breaches as do the dolphins. It feeds on fish, such as herring, sole and whiting as well as other marine animals, such as crustaceans and cuttlefish.

Killer Whale *Orcinus orca*
Known also as the Grampus, (but note that *Grampus* is also the generic name of Risso's dolphin), this species is found throughout the oceans of the world and is fairly common in Irish

waters. It has a powerfully-built appearance, and whereas the male reaches a length of about 9m, the female is only half as big. A beak is absent and the forehead recedes. Its dorsal fin is situated in the middle of the back. In the males the fin is tall, being about 1.8m in height and triangular; in the females it is of moderate size with a concave hind margin. The back is black and the belly white, but above the eye there is a long, oval, white patch which is repeated below, and the white of the belly extends upwards and backwards from level with the dorsal fin towards the flukes. The black flippers are more or less rounded and disproportionately large in old males, and the flukes are black above and white below. The Killer whale has ten to fifteen large teeth on each side of its lower and upper jaws. This most aggressive whale hunts in packs, feeding on seals, sea-lions, porpoises and other whales, as well as fish, including salmon.

Pilot Whale *Globicephala melaena*
This relatively common cetacean is known also as the Caa'ing whale or Blackfish. The female is about 6m in length and the male up to 8.5m. They are long and slender in appearance but above the inconspicuous beak the head is protuberant. The dorsal fin, situated in front of the middle of the back, has a long base but little height. The flippers are long and tapering, being about one-fifth of its body length. Except for white at the throat, it is black all over. Feeding on cuttlefish, it has eight to ten teeth on each side of the upper and lower jaws. This species is gregarious, many hundreds having been recorded

in a single school. In the Faeroe Islands, when a school is sighted, it is driven into a harbour and the whales slaughtered according to ancient rules. The herding instinct of these whales makes this possible.

Risso's Dolphin *Grampus griseus*
This dolphin is distributed throughout the world and has been stranded at least eleven times since 1913 on Irish south and west coasts. About 4m in length, it has a prominent forehead which recedes only slightly from the end of the upper jaw, and it is beakless. The dorsal fin is situated in the middle of the back and the apex points backwards. The flippers are one-sixth of its body length and they taper to a point. Grey on the back with the head lighter in colour, the sides may be darker but the belly is white. The fin, flippers and flukes are black. Generally having four teeth on each side of the lower jaw, it has no teeth in the upper jaw.

White-beaked Dolphin
Lagenorhynchus albirostris
A North Atlantic and Baltic species, this is said to be one of the commoner species. Its length is about 3m and it possesses a distinct beak about 5cm long. The dorsal fin is conspicuous and the tip of the apex turns sharply backwards. The flippers are inserted broadly but taper to the rounded apex. The beak is white. As far back as the rear of the dorsal fin the back is black, as is the head, but there are pale areas on the sides which sometimes join behind the dorsal fin. A

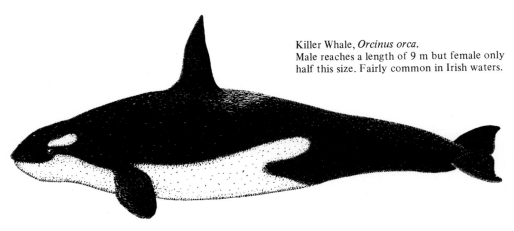

Killer Whale, *Orcinus orca*.
Male reaches a length of 9 m but female only half this size. Fairly common in Irish waters.

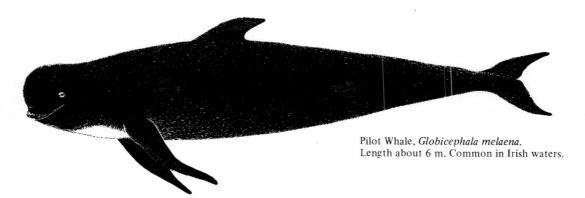

Pilot Whale, *Globicephala melaena*.
Length about 6 m. Common in Irish waters.

dark area behind the flippers extends down-wards almost to the white belly, which changes colour at the vent. A dark line runs from the flippers to the mouth. There are from 22 to 25 teeth on each side of the upper and lower jaws. The White-beaked dolphin sometimes occurs in very large schools which feed on herring, cod and whiting. There have been seven recorded occurrences on the Irish coast between 1851 and 1931 but there has been none since.

White-sided Dolphin
Lagenorhynchus acutus
This rather robust-looking dolphin is of North Atlantic distribution, but has occurred, from time to time, as far south as Kerry. Up to 3m in length, the general form of the body is similar to that of the White-beaked dolphin. The pigmen-tation is distinct in that the flippers are not inserted in a dark area, and there is a wide, long, light-coloured area on the sides from the level of the dorsal fin backwards. There is also a greater number of teeth — from 30 to 34 on each side of the upper and lower jaws. This dolphin occurs in large schools but is not as abundant as the White-beaked dolphin.

Bottle-nosed Dolphin *Tursiops truncatus*
Almost as abundant as the Common porpoise, the Bottle-nosed dolphin is frequently seen in the south-western approaches to Ireland and Britain. Its shape is robust and it is up to 3.7m in length with a pronounced beak 6 to 7cm long. The apex of the dorsal fin points backwards. Its back, flippers and flukes, as well as the ventral surface of the tail region are dark brown or black but the

throat and belly are white. There are from 22 to 25 teeth on each side of the upper and lower jaws. This dolphin is usually seen in small schools, and it is a delightful sight to watch them swimming in front of a ship as they weave their way in and out — the young ones following their mothers' every twist and turn. From time to time they are known to ascend rivers for some distance from the open sea. They are fish and squid eaters.

The dolphin has played a significant part in fables and legends since the earliest times. Although a predator on schools of fish and potentially dangerous, it has always displayed a remarkable friendship with man, and there is no known instance of it attacking humans. Old stories of children being saved from drowning by dolphins carrying them on their back are paralleled by many modern instances of their playful habits with swimmers. Dolphins communicate with each other by a series of whistling and cackling noises, and they are able to click their tongues and send out cries of distress, but it is only in recent years that their remarkable intelligence has been fully appreciated. Dolphins have been taught to imitate the human voice; to repeat words; to whistle and laugh and to carry out exercises in salvage and naval operations.

Common Dolphin *Delphinus delphis*
The Common dolphin is a creature of superb grace in its marine element which makes it familiar to coast-dwelling and sea-going people. It is identified by its slender shape, its deeply-indented dorsal fin and projecting beak, about 15cm long, and its white flank with longitudinal

39

above: Common Dolphin, *Delphinus delphis.*
Usually about 2 m in length and common in Irish waters.

below: Common Seal, *Phoca vitulina.*
Length of male up to almost 2 m, female considerably
smaller. Distributed around the coast and
generally associated with sandbanks and estuaries.

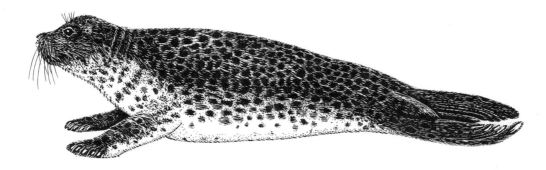

below: Grey Seal, *Halichoerus grypus.* Length of bull
2.3 m, cow just over 2 m. Common all around the coast.

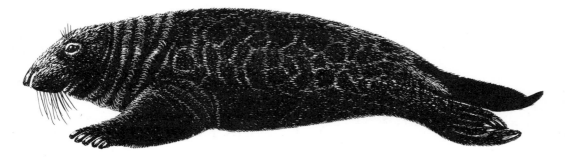

dark stripes. Varying in colour from black to brownish-green, it is usually about 2m in length but may occasionally reach 2.6m, and weighs between 70-75kg. There are 40 to 50 teeth, 2.5mm in diameter, on each side of the top and bottom jaws and they curve slightly backwards. They are so spaced that they fit between each other when the jaw is closed.

The Common dolphin is distributed throughout the tropical and temperate seas of the world, but it may occur in more northerly regions. It is gregarious, a herd consisting of from ten to about 100 individuals, and being a sociable creature, will aid a sick or wounded member of the herd by supporting it so that it can breathe. The dolphins' habit of accompanying ships is well-known but, in the writer's experience, it is more usual for the Bottle-nosed dolphin, *Tursiops truncatus* to do this. Dolphins frequently precede a ship, seemingly taking advantage of the pressure-wave in front of the bows. Here the vertical movement of the tail flukes (for propulsion) may be seen.

The Common dolphin, although mostly keeping to the open sea, will also enter estuaries; that of the Shannon is commonly frequented. One of the most spectacular habits of this species, however, is for it to proceed rapidly across the sea in a series of jumps, hurling itself clear of the water, and the writer has observed this when about sixty individuals were involved.

Seals *Pinnipedia*

Seals were formerly classified as a division of the *Carnivora,* but are now usually included as a separate order, together with the Sea-lions, Eared seals and Fur seals — the *pinnipedia.* The two Irish species of seal belong to the *Phocidae,* the True Seals, or Earless Seals. They are highly adapted for their aquatic environment.

The seals are characterized by their body being streamlined and covered with short, coarse hair. The neck region is inconspicuous. Limbs are modified as flippers, the digits not being separate although all bear claws, with those on the forelimbs more pronounced. When moving about on land, only the forelimbs are of use, the hind-limbs being dragged behind; but in water seals are fast and very agile.

Common Seal *Phoca vitulina*

Also known as the Harbour seal, this species, outside the breeding season, is generally distributed around the Irish coast. It is associated with sandbanks and estuaries; although in Northern Ireland it hauls out onto rocks (not entering small bays — where a good look-out cannot be kept.)

The Common seal is a dark grey colour on the back, paler on the belly but with an approximate uniform mottling with abundant small, black spots. The nostrils are located on the muzzle at an angle of about 45°, the muzzle being broad and blunt with the top concave (*retroussé* nose). The length of the male — head to tip of flipper — is from 153 to 198cm and the female 137 to 168cm. The weight of the male varies from 202 to 253 kg and the female 100 to 150kg. Little is known of their breeding habits but there is no defence of territory by the male. Mating takes place by the water's edge or under water from a depth of 4 to 8m. There is a period of delayed implantation.

The young are born during June, in the sea or on sandbanks and rocks covered at high tide. This means that the newly born seal must swim within a few hours of birth. Suckling takes place under water at first and lasts for three or four weeks. Food consists of crustacea and molluscs with the principal fish consumed appearing to be flatfish.

Grey Seal *Halichoerus grypus*

The Grey seal is generally more northern in distribution than the Common seal, although Ireland is entirely within the range of both species.[4] It is common all around the Irish coastline and even though it prefers breeding sites facing the open Atlantic, it uses caves on Lambay Island less than fifteen miles from Dublin! It is sometimes referred to as the Atlantic seal, and is associated with rocky coasts and inaccessible islands.

The coat-colour of this seal is very variable. Bulls are darker than cows, from black to mid-grey or mid-brown with small, paler spots or areas. Cows are greyish to fawn and much lighter underneath and spotted more or less overall. The head profile is convex (Roman nose) in the bulls and straight in the cows, and the nostril slits are near vertical. The two latter characters are the only reliable ones in the field, colour and spotting being useless for identification.

The adult bull has an exceptionally thick neck usually heavily creased. An accumulation of blubber enables him to starve for about two months around the breeding season — the cow starves for about two weeks only. The length of bulls from head to flipper averages 232cm and cows 204cm.

Breeding commences in late August when bulls come onto the site and establish territory. The pups are generally born about September and October. Males are not sexually mature until six years old but competition for territories prevents them from mating until much later. The cows come into season two or three weeks after birth of a pup, but delayed implantation takes place, so that total gestation is for 11½ months. Active embryonic growth takes about 7½ months. Only a single pup is produced each year. Cows live for up to 34 years and bulls up to 23 years. Food consists of medium-sized crustaceans, such as crabs and lobsters, and fish, such as conger-eel, lump-sucker, saithe, pollack, mackerel, pilchard, herring and salmon. Squid and other cephalopods are also taken. The Grey seal is much persecuted by fishermen, not only because of their consumption of fish and damage to nets, but also because of the value of their skins and oil from the carcase.

NOTES

1. Ninety-three were captured by whaling stations between 1908 and 1914.

2. Two were captured by Irish whaling companies between 1908 and 1914.

3. Forty-eight were taken by whaling stations between 1909 and 1914, and there were eight definite strandings from 1913 to 1974.

4. It is generally more common than the Common seal in Irish waters.

3 Birds

BIRDS *Aves*

Birds are warm-blooded, egg-laying vertebrates specially adapted for flying. The anatomical adaptations and the special skin outgrowths (feathers) are such that a bird can be easily recognised. Among the 8,600 living bird species in the world today there is a basic uniformity, so that even those which have lost the power of flight can still be identified as birds. Possessing two pairs of limbs, the basic bone-pattern is similar to that of mammals although the front pair are modified as wings. This causes all birds to be bipedal whilst the function of the hands is transferred to the beak. The latter is hard, horny and scale-like, and the reptilian ancestry is further shown by the scales which cover the toes and lower part of the legs. Skeletal adaptations contributing to flight are the fusing of certain bones, acting as a fulcrum for the wing action; the development of a keel on the broad breastbone for flight-muscle attachment, and the presence of air-sacs in the bones in order to lighten them. Feathers serve to streamline the general body-shape for flight as well as to conserve heat. Certain feathers on the wings (flight-feathers), extend the effective area of the wings whilst tail-feathers help to stabilize and guide.

The official list of Irish birds gives 375 species. Many of these have only been recorded on a few occasions and, indeed, more than half the list is of uncommon visitors. Ireland is strategically placed to be a landfall for many birds blown off their migration course or swept by storm from the Iberian peninsula and North America. *The Atlas of Breeding Birds in Britain and Ireland* (1976) gives 141 species as breeding in Ireland although, in a few cases, they have either not been confirmed or breeding takes place in only very few localities.

Some birds, the summer residents, spend the summer in Ireland and breed, but then go south for the winter. The Swallow, *Hirundo rustica*, is an example. Others — the winter residents — spend the summer in the north or east where they breed and then spend the winter with us. The Fieldfare, *Turdus pilaris*, is an example. Passage migrants are those which fly through the country during their migration, only staying for a few days or a week to feed and rest. There are also birds classed as permanent residents, with us throughout the year, and of course they are included in the list of Irish breeding birds. Additionally some birds come within a combination of two or more categories. Some species can be both summer and winter visitors, and passage migrants as well as permanent residents.

Notes and illustrations can only be given for a few birds. For an inexpensive book on Irish birds, the reader is directed to Christopher Moriarty's *Guide to Irish Birds*.

Great Crested Grebe *Podiceps cristatus*

The Great Crested Grebe, a bird of our reed-girt lakes, is a species whose numbers have fluctuated widely. In the second half of the nineteenth century it was hunted relentlessly for 'fashion' feathers, mostly for ladies' muffs which were made from the feathers, down and skin of its breast, and was almost wiped out in Ireland.

Since the turn of the century, however, the population has increased to a great extent. It was with the passing of the fashion using its feathers, when Acts of Parliament made it unlawful to kill birds for their plumage, that the Great Crested grebe became again a beautiful adornment to our lakes. Except for the north-west, it is a common breeding bird in all suitable localities in the northern half of Ireland, and in the south, Co. Wexford was colonized in 1946 and Co. Cork in 1967.

The Grebes go through courtship displays which are ritual and often elaborate. The so-called 'penguin dance' consists of both birds diving, coming up with weeds in their beaks and then rising breast to breast with beaks extended. The four or five eggs, which are laid in May, are elongate-oval in shape and at first chalky-white. They rapidly become stained and just before hatching may be dark brown or even red, when ferrous oxide is present. Incubation takes about four weeks, the eggs being covered with water weeds when they are left by both parents. The chicks are fed by the parents for about two months, and are then independent.

Cormorant *Phalacrocorax carbo*

The Cormorant is a large, almost black aquatic bird, generally marine and coastal but also breeding (although uncommonly) on inland waters. It is seldom observed out of sight of land. The beak is fairly long and hooked at the tip, and there is a white patch covering the cheeks and chin, and it is this which immediately distinguishes the Cormorant from the only other species, the Shag, *Phalacrocorax aristotelis,* with which it might be confused. (Additionally the latter possesses a more slender beak and is generally smaller and more lightly built and crested in spring). The neck of the Cormorant is long and rather snake-like, blue-black in colour. The wings are dark, metallic bronze and the legs are stout and the feet webbed. In the breeding season there is a conspicuous white patch on the thighs.

The Cormorant is often seen in flight with outstretched neck, keeping only one or two metres from the sea surface, and it is often seen to fly a considerable distance in a perfectly straight line. The wing beats are rapid but there are occasional glides. When flying overland it does so at a greater height and it is known to soar at considerable altitudes. Its food is obtained by diving and consists mainly of fish which are brought to the surface for swallowing. The dive may be initiated by a jump upwards from the swimming position, or by a gentle subsidence with hardly a ripple. Propulsion under water is by simultaneous strokes of both feet with wings closed, although they can be opened a little as a

Great Crested Grebe, *Podiceps cristatus.* Total length 47.5 cm of which body is 30 cm. Common resident north of a line from Dundalk to Limerick. Local elsewhere but not increasing.

Cormorant, *Phalacrocorax carbo*. Total length
90 cm of which body is 55 cm. Resident and partial
migrant. Generally distributed around coast and
frequently occurring inland.

brake. After an active period of fishing the bird
takes up a resting position near the water's edge,
when it will characteristically hold its large wings
out to dry. There has not been a satisfactory
explanation for such behaviour except, rather
obviously, the bird believes that they have
become wetted!

Breeding places are well advertised by the
extensive patches of white guano and are usually
found on broad ledges on cliffs, on the flat tops
of inaccessible rocks or small islands. It is a
colonial nester, with sometimes a few, but up to
150 nests being located within less than a third
of a metre of each other. In addition, nesting
colonies exist in trees in some inland areas but
these appear to be less numerous than once was
the case. The three or four eggs are laid in late
April to early May on a nest of seaweed or
heather stalks (inland), lined with grass, rushes
or straw. The eggs are pale blue which is almost
obliterated by a white, chalky deposit. Their
shape is given as 'elongate-oval', rather blunted.
Incubation takes 28 days and after hatching the
young are fed by both parents with partially
digested food which the young take from the
parent's mouth. They are fledged in about a
month. In Ireland breeding takes place in suitable
places around the coast except in the east where
considerable stretches of coast-line are devoid of
suitable nesting places.

Tufted Duck *Aythya fuligula*

The Tufted Duck is a recent coloniser of our
lakes, lochs and reservoirs. Indeed, the last few
years have seen a phenomenal population
increase of this dapper little black and white
species, generally in freshwater habitats. At
least, the drakes are black with white flanks and
a backwards-drooping crest. Where he is shining
black, she is a dark brown colour with paler
brown flanks.

Tufted Duck, *Aythya fuligula*. Total length 42.5 cm
of which body is 27.5 cm. Resident and
Winter visitor. Breeds in all counties except
Waterford and Carlow. Increasing in abundance.

In winter, large sheets of water may hold hundreds, if not thousands, of Tufted Duck, even where there is no marginal, protective vegetation. But it is as a breeding bird that it now has a major significance in Ireland. It now appears that its present breeding distribution is limited by the availability of suitable areas of water, which may be man-made (such as gravel pits), as each pair seems to require at least 1 ha of water for breeding to take place. The largest known nesting areas are on Loughs Beg and Neagh where there are approximately 1,000 breeding pairs. There are 200 breeding pairs on Lower Lough Erne in Co. Fermanagh. Even so, the almost explosive expansion of this little duck may not yet have reached its ultimate distribution. Breeding in Wexford and Waterford commenced as recently as 1970.

The food of the Tufted Duck consists of aquatic snails, insects and other invertebrates which are hunted for by diving and searching the lake bottom, but some aquatic vegetation is also eaten. It usually dives in depths of 1 to 2m, although it has been known to dive to 5m and to turn stones over underwater. The nests are generally hidden in dense vegetation on land within 10m of the water's edge, and sometimes advantage is taken of an old Coot's nest floating on the water. In some situations the Tufted Duck has taken up the social nesting habit — several hundred nests located close together, often with other duck species or gulls. It is a late nester, mid-May being the usual date for the beginning of laying, and up to 14 eggs have been recorded from a single nest, but sometimes not all laid by a single duck. The eggs are comparatively large and matt, greenish-grey in colour. They hatch in just over three weeks, and after only a few hours the ducklings can dive and at six weeks are capable of flight!

Canada Goose *Branta canadensis*

Only two species of geese breed in Ireland — the Greylag, *Anser anser,* and the Canada goose. The former is a feral population and its main nesting area is around Strangford Lough in Co. Down, although this species probably bred in Ireland 300 years ago. The Canada goose is derived from North America. It has been known in Britain as a decorative water bird for lakes and pools in the grounds of gentlemen's residences and the parks of the nobility since 1676, and there have been many introductions during the intervening years. Although in Britain there have been semi-feral flocks since 1840, it is only since 1950 that much wider distribution has taken place — almost certainly helped by transportation by wildfowl interests. As a bird for sport, however, it did not live up to expectations. They are often tame, with no flighting pattern and, indeed, seldom reach more than tree-top height when flying.

In Ireland, there are three important full-winged flocks but all associated with wildfowl collections. They are in Cork, Fermanagh and especially County Down. Canada geese in the last-named county nest around Strangford Lough and are now considered truly feral. They can be identified by their size, for they are large, up-standing birds with pale brown belly feathers and cream-edged, brown feathers on the back. Most of all, however, it can be distinguished from all other geese by the long, black neck and head which has a white patch stretching from the chin to just behind the eye.

Three other species of geese occur in Ireland as winter visitors. These are the Greenland White Front, *Anser albifrons flavirostris,* of which there are usually about 8,500 — the greater number wintering in the North Slob, Co. Wexford. The Barnacle, *Branta leucopsis,* is chiefly found on islands of the west coast although it is recorded from a number of other localities, some inland. About 4,500 in all winter in Ireland, and they also breed in Greenland. The numbers of the Pale-breasted Brent, *Branta bernicla lirota,* have increased in recent years from nine to 14,000. They are to be seen in a number of localities all around the coast.

Mute Swan *Cygnus olor*

The Mute Swan is Ireland's largest breeding bird and its white plumage and arched neck must be generally familiar to everyone. It has a strange history. Probably introduced by the Normans, its ownership became vested in the Sovereign in mediaeval times. As well as being a most beautiful ornament to any patch of water, swans were

right: Bewick's Swan, *Cygnus bewickii*

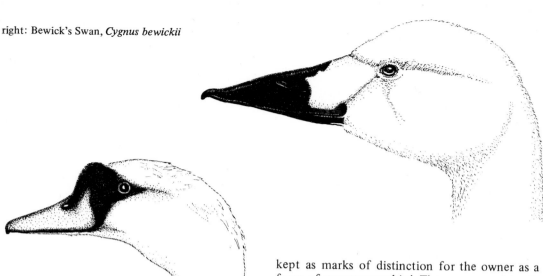

above: Mute Swan, *Cygnus olor.*
Total length about 1.5 m of which
body takes up about one half. Common
resident, breeding in every county.
Increasing in abundance.

below: Canada Goose, *Branta canadensis.*
Total length from 80-100 cm of which body
is from 52.5-60 cm. Annual vagrant with
feral populations associated with waterfowl
collections. Is rare as a vagrant.

kept as marks of distinction for the owner as a
form of swanupmanship! They were used tradi-
tionally as gifts between dignatories and were
valuable items on the menu at banquets.

Two other swan species are winter visitors to
Ireland — the Whooper, *Cygnus cygnus,* most of
which breed in Iceland, and Bewick's Swan,
Cygnus bewickii. I well remember, many years
ago, sighting my first herd of Whoopers — twelve,
large birds with long, straight necks ending in
large, yellow beaks without a black knob at their
bases, stood well out from the shore of Lough
Erne. The thrill lasts to this day!

Sparrowhawk *Accipiter nisus*

The Sparrowhawk is a relatively common bird
throughout the whole of Ireland wherever there
is sufficient cover for its principal prey of small
insectivorous and seed-eating birds. The females,
which are larger than the males, regularly kill
wood pigeons, *Columba palumbus.* The writer
recalls most vividly one such occasion when a
Sparrowhawk suddenly alighted in front of him,
laying aside its habitual caution, with a starling
clasped in its talons which, seconds before, had
been snatched out of the sky! Although not a
muscle twitched nor an eyelash flickered, the
hawk, after a minute's uneasiness, looking this
way and that, suddenly shot off.

The Sparrowhawk works on the hedges and
edges of woodland, gliding swiftly to one side,
then, slipping through an opening with a few,
quick wingbeats, searches along the other. Small

47

Sparrowhawk, *Accipter nisus*. Total length about 27-37 cm.
Female considerably larger than male. Fairly common
resident breeding in every county. After decreasing in recent
years now appears to be recovering.

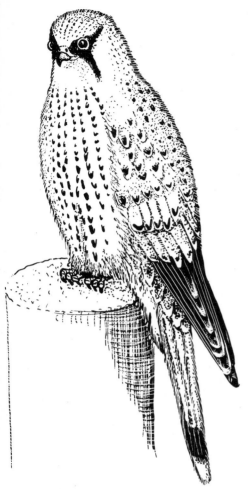

Kestrel, *Falco tinnunculus*.
Total length from 32-35 cm. Sexes
not very different in size.
Widespread resident breeding
in every county.

The characteristic way it hunts its prey is, perhaps, the easiest method of identification, but the broad and rounded wings and the generally barred underparts (including the tail), should differentiate this species from all other birds of prey. Males and females are very distinct. Whereas the upperparts of the male are dark, slate-grey with the underpart barrings reddish, the much larger female has the upper parts more brownish and the underparts whitish, but there is much individual variation. Apart from the preying flights, a number of types of display and courtship flights have been described, including soaring to high elevations. In mixed woods, the nest is frequently constructed in a spruce, but in deciduous woodland is likely to be built near the main trunk of an oak at some distance from the ground. Often the female builds it entirely, but the male occasionally brings along larch twigs and other material. Bark, decayed wood or dead leaves are used as a lining. Coition is in the nest. Generally 4 or 5 eggs are laid, but much higher numbers are due to the efforts of *two* females. Their ground colour is usually dirty-bluish or greenish-white with blotches, spots or streaks of dark, chocolate brown. A full clutch is laid by the end of May and incubation takes five weeks. After hatching the male brings food either to the nest or to the 'plucking place' where the female apportions it out. Air transfer of prey may also take place. Fledging can take as little as 21 days or as long as 30 days.

Kestrel *Falco tinnunculus*

The Kestrel is a common and widely distributed falcon in Ireland. Its characteristic habit of hovering; facing the wind poised in one spot 10m or so above the ground; its wings alternately fluttering or held still; its tail twisting or fanning, must have been seen by everyone in the countryside or by anyone who had eyes about him or her. If one is interested in nature, one cannot help the eyes becoming glued to the hovering bird — to watch it slip down a few metres, hover again, and then perhaps make a slanting dive onto its prey. When the bird is perched on a fence or a telegraph post we may then be unsure of its identity, sometimes confusing it with a Sparrowhawk, if an inexpert observer.

birds taking insects in the forest canopy are suddenly made to dive into cover of brambles, or blackbirds seek safety in deep shrubberies — sometimes quite close to buildings. Indeed, Sparrowhawks are known to frequent even the centre of large cities if meals are in the offing, and the great masses of Pied Wagtails congregating in O'Connell Street, Dublin, for communal roosting, are regularly culled by Sparrowhawks.

The male Kestrel, however, should be differentiated quite easily with its bluish-grey head, rump and tail and its chestnut upper parts richly spotted with black. The female, on the other hand, is not so distinctive, the upper parts being rufus brown with blackish or very dark barrings. In both sexes the lighter underparts are streaked. In the Sparrowhawk the underparts are barred, not streaked, and whereas the male has slate-grey upper parts, those of the female are brown. In flight the Kestrel has pointed wings whilst those of the Sparrowhawk are rounded.

The Kestrel is not a bird of thickly-wooded areas, but of rough grassland and hillsides. This is why it is seen so commonly hovering along the gently sloping cuttings and embankments of new motor roads. It is the tussocky herbage that gives shelter to fieldmice, but their movements do not escape the sharp eyes of the hunting Kestrel. It eats many beetles and occasionally small birds.

The Atlas of Breeding Birds shows that there are few 10km squares throughout Ireland where it does not breed. There has been a long period of build up in numbers but it has been stated that it might now be decreasing. No nest is made, but the 4 or 5 eggs are laid in April or May in old nests of crows, magpies and other species and on cliff-ledges or in hollow tree trunks. The white eggs are washed and blotched with dark, red-brown so that some, and in many cases, all, of the white, ground colour is lost. They hatch in about four weeks having been incubated mainly by the female. The male catches all food during this period and transfers it to his mate.

Hen Harrier *Circus cyaneus*

The Hen Harrier is probably the most numerous of the larger birds of prey in Ireland, considerably outnumbering the buzzard. The latter is to be found only as a breeding bird in the north-east and rarely in the north-west. The head of the Hen Harrier appears large and squat due to the owl-like facial disc. The yellow legs are long and thin, but the large black claws are prominent. The male is bluish-grey with a white belly but its most characteristic feature is the most startlingly white rump. This is to be seen in Co. Cork when viewing the most southerly breeding Hen

Harriers. They hunt the fern-lined streams adjacent to sedgey meadows in this area, where new forestry plantations grow on the hills giving protection, not only to a number of prey species but also to the harriers themselves, as the degree of human molestation decreases. It is precisely in this type of habitat that the Hen harrier has increased in numbers, having been almost wiped out when game bird preservation was a very serious business. Since 1950 they have made something of a comeback, as by 1964 they were breeding in six counties and thirteen by 1971. In the early 1970s there were probably between 200 and 300 pairs throughout Ireland.

The harrier hunts at a height of about 2m, alternately gliding and flapping, quartering the ground for small mammals and birds. When a luckless prey is sighted, it is dropped on in an instant, being taken completely by surprise.

Except for the Counties of Kerry and Cork, the Hen Harrier is generally distributed over the southern half of Ireland. It occurs also in the central areas of the north but is rare or absent from the north-west and the north-east.

The nest consists only of a hollow in the ground which the female lines thickly with grass and rushes. Four or five eggs are laid in early May and are bluish-white in colour and almost always unmarked but occasionally have light, rusty-red spots or, more rarely, streaks of the same colour. They hatch in about one month and six weeks or so later the young are fully fledged. During the incubation period the female is 'called' from the nest by the male who passes food to her in the air; either directly, foot to foot, or by dropping it for her to catch.

Corncrake *Crex crex*

The Corncrake is a common but fast-decreasing Irish bird in spring and summer, and is seldom seen by the casual observer. The grating voice of the male, however, which is represented by its scientific name *Crex crex* and often monotonously repeated, is one of the most well-known sounds of the Irish countryside. The well-nigh invisible birds skulk amongst the mowing grass and the damp herbage of low-lying fields and rushy meadows. When disturbed it seldom flies far, and may be easily identified by its chestnut-

Hen Harrier, *Circus cyaneus*.
Total length from 42.5-50 cm.
Female larger than male.
This resident was once on the
verge of extinction but is
now plentiful in some of the
13 counties in which it
now breeds.

above: Woodcock, *Scolopax
rusticola*. Total length about 34 cm
of which the bill is 7.5 cm.
Common resident and Winter
visitor breeding in every county.
There is some migration to Europe.

left: Snipe, *Gallinago
gallinago*. Total length 26 cm.
A common resident and
Winter visitor in great numbers.
Breeds in every county.

above: Corncrake, *Crex crex*. Total length about
26 cm. Summer visitor, occasional Winter visitor and
passage migrant. Decreasing in numbers except
in the North-west.

coloured wings; more usually it seeks shelter by running away. When seeking its mainly insect food it walks rather like a moorhen, moving its neck backwards and forwards and flicking its tail upwards. At times it flies more strongly than is usual but not at much height, and frequently falls as a casualty after a collision with overhead cables. In spite of its generally weak flight, however, it is a summer visitor and migrates to Africa for the winter.

The nest consists only of a pad of dried grass amongst long herbage, and the eggs, which may number up to 14, are laid in May. They are pale, greenish-grey to light, reddish-brown and freely spotted with red-brown and grey. Incubation takes only two weeks and the young are independent after a few days, flying after about four weeks. The Corncrake population has decreased considerably during this century, and this decline has been put down to machine cutting of corn and grass destroying nests. As a breeding bird in Ireland, it is known almost throughout the country except in some areas of Cork, Waterford and Wexford.

Snipe *Gallinago gallinago*

The Snipe is a typical bird of much of the Irish landscape. It usually inhabits wet areas such as poorly-drained and rushy fields, water meadows, bogs, saltmarshes, edges of rivers and lakes where they are marshy — in fact any muddy place where there is little disturbance. Indeed, there are very few areas of Ireland where it not known as a breeding bird. Because of its habit of hiding during the daytime it is not often seen, and is flushed only when about to be trodden on! Then, uttering a hoarse cry, it flies up in a series of zig-zags until it is clear of the intruder. The bill is long and straight, being used for probing the mud right down to the base, making jabbing movements. When a worm is found it is swallowed without withdrawing it. Its plumage is richly barred and patterned in rufus browns and buffs, and the tail is barred with black and tawny. The flanks shade from white to buff, and are more or less distinctly barred.

When Snipe are present, their detection is most often caused by sound, especially when they are active. Usually in rain or at dusk in the breeding season they utter their well-known 'chip-per', 'chip-per' or 'jick-jack', 'jick-jack'. Then, more so on moonlit nights, but also on warm summer evenings, they perform their aerial display known as 'drumming'. The bird flies rapidly up to a height from 20m to 100m, then dives down at great speed at an angle of about 60°.

The wings are half-closed and quivering and the tail spread out with the outer pair of feathers standing out. This causes a resonant sound which lasts up to about three seconds, but the performance, carried out by both sexes, may be repeated many times — by night as well as by day. The 'drumming' sound is caused entirely by the vibration of the outer tail feathers, but for many years it was the subject of much controversy. Snipe also perform a number of aerial evolutions including gliding on their backs.

The nest, which is a hollow lined with dry grass, is situated in marshy pastures, usually in a grass tussock or clump of rushes. Usually four eggs are laid in early April and onwards, and are olive-grey to olive-brown in ground colour with bold blotchings and spots of sepia and Vandyke brown, and some almost black. There is a wide variation, however, and eggs pale blue to deep amber are recorded. The female incubates for twenty days and the young leave the nest as soon as they are dry, attended by both parents who divide the brood between them and go their separate ways. The downy chicks are said to be the most attractive of all waders in their russet- and silver-spangled colouring. They fly at about three weeks.

Woodcock *Scolopax rusticola*

The Woodcock is a well-known breeding bird in Ireland and is well distributed except for an area taking in Kerry and Cork. In this regard, the pattern resembles south-west Wales and the English counties of Devon and Cornwall. In addition to the breeding population, large numbers are immigrants from England, Scotland, Scandinavia, Russia and Germany but many of these fall to the gun.

The Woodcock is a wading bird, but a wader of the forest, and its beautiful, 'dead-bracken' camouflage makes it virtually invisible when sitting on its nest in the forest undergrowth. It uses its long, thin bill with its sensitive tip to dibble worms from the mud at dawn and dusk, so that altogether it is a most secretive bird. However, its curious flighting in the breeding season, known as 'roding' may often be watched as both sexes fly around their territory at about tree-top height. The same path is kept with remarkable accuracy, and from time to time a strange 'squeak and groan' is made by the momentary cessation of wing beats.

The Woodcock is the size of a Wood Pigeon and it has a uniquely-shaped head which slopes steeply backwards from the beak, and the eyes are located high up. Its nest is sited on mossy ground in woodland and is lined with leaves. The eggs are usually four in number and are from greyish-white to a warm-brown, splashed with various shades of chestnut, and ashy-grey in colour. It is double-brooded and the adult birds are able to carry the young in flight held between the thighs.

Curlew, *Numenius arquata*. Total length averages 66 cm. Common and widespread resident and Winter visitor. Breeds in every county.

Curlew *Numenius arquata*

This large, wading bird with a long, curved beak is perhaps best known for the sounds it makes, exemplified by its name. The 'cur' is drawn out and the 'lew' short and at a higher pitch. In addition, its bubbling call at breeding time is a nostalgic sound reminiscent of upper moorland. Within recent years there has been a significant increase in the numbers of Curlew. Instead of breeding at high altitudes on damp and boggy moors, it now nests in many lowland areas where there are open, damp pastures and other poorly drained regions. It is now known as a breeding bird throughout Ireland but is not so abundant along the south coast strip.

The adult bird is up to 60cm in length, including the beak, which is about 12cm long. The colour of the plumage is best described as streaky-brown with the belly white and the tail barred. It is difficult to approach.

Its nest consists of a hollow on the ground lined with grasses and sheltered by vegetation to a lesser or greater extent. The four eggs are laid either late in April or early in May and are somewhat pear-shaped. Their ground colour varies but is usually greenish, olive, brownish

to dark reddish-brown, irregularly marked with sepia, umber and chocolate. Incubation takes between 29-30 days and is by both sexes. The young leave the nest as soon as they are dry, being tended by both parents for five or six weeks when they can fly. In July a high proportion of the birds leave for the sea shores and estuaries where they feed on small marine animals. Flocks also make regular excursions to cultivated land where they feed on insects and other invertebrates.

Common Sandpiper *Tringa hypoleucos*

The Common Sandpiper is a bird of stony-edged streams and rivers, lakesides and, in the south-west, shingly shores of bays and estuaries. As a breeding bird it occurs now mainly in the West, from Donegal to West Cork, but in the first half of the present century its range contracted after being abundant in Carlow, Kilkenny, Waterford, Wexford, Kildare, Louth, Meath and East Cork. It is now scarce in these counties.

These lark-sized wading birds are delightful to watch as they make short, tripping runs along the water's edge. They bob, showing their white, under-tail coverts and they frequently call 'willy-wicket'. Short flights are made over the water and their white bars along the wings flash, and they 'pipe' as they go. It is the only common

Common Sandpiper, *Tringa hypoleucos*. Total length about 19 cm. Widespread Summer visitor but scarce in South-east and breeds in every county except those of the South-east.

Irish wader which is a summer visitor. The nest is a hollow in the ground on a shingle bank or the lakeside, and sheltered by a tuft of vegetation. The four eggs are usually creamy-buff, spotted with dark, chestnut-brown and ashy-grey. It is single-brooded.

Little Tern *Sterna albifrons*
This is Ireland's rarest, regular breeding bird. Between 1967 and 1971, the number of breeding pairs varied between 110 and 296. The distribution of the colonies, or sometimes single nests, is discontinuous, stretching from Lough Foyle to Galway Bay along the north and west, and from Drogheda to Cobh in the east and south. In 1948, the Lough Foyle colony consisted of only one pair — I nearly trod on the nest and the parents fluttered around me like butterflies as they escorted me away! The scrape which served as the nest was very exposed and perilously near high tide mark, but this appears to be a common practice with the species. This means that the Little Tern faces three main hazards — sand, water and people.

Collared Dove *Streptopelia decaocto*
The Collared Dove is now found throughout Ireland wherever there are habitats around farm buildings where some spillage of grain or cattle feedstuffs occurs. It is even to be found in towns where there are granaries. Its distribution throughout the British Isles is nothing less than explosive! It was not until 1955 that the first of the species nested in Britain, but by 1960 they were already breeding in the districts around Dublin and Belfast. If we go back to 1930, they were found in Europe only in Turkey and some areas of the Balkans. Czechoslovakia was reached in 1936; Austria in 1938; Germany in 1943, and they bred in France in 1950 and in Norway in 1952. They now breed in Ireland.

Its plumage is beautifully toned in soft-grey and fawn, with black wing-tips and a black collar at the back of the neck. It seems a great pity that some look upon it as a pest. I find it wholly delightful to watch its little flights around the house as it planes and flutters, and I have friends who find its voice monotonous, but I do not find it so. The male's call as it flies onto a perch is attractive to my ears. It is a prolific breeder (as might be suspected) — brood following brood throughout the summer. Ivy covered walls, conifers and holly are preferred nesting sites for its flimsy nest.

Swift *Apus apus*
It is seldom that we pause to marvel at the great mastery of aerial flight possessed by the Swift. It is well-named, as a chasing group come wheeling out of the sky with a crescendo of screeching whistles, diving and gliding; then, with a flutter of rapid wing beats, they are off again. With what

Little Tern, *Sterna albifrons*. Total length about 24 cm. Summer visitor breeding in small colonies on the East, North-west and West coasts. Maybe decreasing in numbers.

Collared Dove, *Streptopelia decaocto.* This resident is wide-spread, although there are areas from which it is absent such as the North Midlands.

Swift, *Apus apus.* Total length about 16 cm. A numerous, well-distributed, Summer visitor increasing in abundance.

sureness and confidence they avoid the roofs and chimneys and other obstacles! Then, when they approach the nest hole under eaves or amongst displaced masonry, they swoop with such speed that they must surely collide with the building. At the last split second, their momentum vanishes and they are clinging to the vertical wall, soon to scuffle into the nest chamber.

The Swift is fairly easy to identify as it can only be confused with the Swallow, *Hirundo rustica,* and the two Martins, the house Martin, *Delichon urbica,* and the sand Martin, *Riparia riparia.* The wings are very different from the three latter species, being narrow, rounded and scythe-like in shape. The tail is short and the general colouration dull, sooty-brown. It is seldom that one gets near enough to make out the more intricate features, but the primary feathers of the wings show a lighter edging, especially in young birds, and the forehead is somewhat frosty in appearance.

Only once in my lifetime have I had the opportunity to examine a Swift when alive. I picked up one under a church steeple in which there was a prodigious number of breeding Swifts. It was a young bird, just fully feathered and as I bent forwards to pick it up it looked at me with eyes full of apprehension. Suddenly,

I recoiled with something akin to horror! It was covered with large parasites which were about the size of a house-fly, but wingless, with distended, greenish abdomens. They crawled crab-like over their luckless host's body. I knew them to be blood-sucking flies of the group PUPIPARA, several species of which infest Swallows, Martins and Swifts. How strange that these beautiful, delicate creatures are preyed upon by these members of the HIPPO-BOSCIDAE family.

Pupiparous flies are divided into three families and are especially well-known to those who handle sheep — the sheep ked belongs to the HIPPOBOSCIDAE and is easily identified because the head is deeply inserted into its thorax and the short, stout proboscis points forwards. Its claws are strong, enabling it to creep easily through feathers, and the whole insect is hairy with vestigial wings or none at all. The fly is flattened, as are many parasitic insects, but in this instance top to bottom, not side to side (as are the fleas). There is a peculiar feature of its life history in that the adult females do not lay eggs but, from time to time, produce a fully-fed larva, which immediately pupates when deposited. This is dark red and almost globular, and the shape enables it to roll to the bottom of the host nest, to be incubated by the warmth of the returning migrant the following May.

You may ask what happened to the Swiftlet? I begged a boot-box from a nearby shop and took the bird home, where I removed its tormentors with the aid of forceps. I then took it into the garden, gently throwing it into the air, but although it glided well it lacked the wing-beating flight necessary to become airborne. I repeated this action a number of times but the pectoral muscles would not work. Finally I took it to an upper window, and as I released it yet again it glided over the lawn into the woods and eventually out of sight — but I did not see those scythe-like wings beating.

The Swift is more closely related to the Nightjar than to Swallows and Martins. Arriving in May, it lays two eggs later in the month which are incubated for about three weeks. It leaves as early as mid-July.

Hooded Crow *Corvus corone cornix*

The all-black Carrion Crow, *Corvus corone corone,* and the grey and black Hooded Crow, *Corvus corone cornix,* are now considered to be different forms of a single species, as indicated by their scientific names. Where populations of both forms co-exist, they freely interbreed and the offspring are fertile; this is one of the prime requirements for the establishment of single species rank. The Hooded and Carrion Crow make up one of the widespread bird species in Ireland and Britain, and the survey of breeding birds completed in 1976 suggests that it is,

Hooded Crow, *Corvus corone cornix.* Total length about 46 cm. A number
numerous and widespread resident breeding in every county and increasing in abundance.

indeed, second only to the Skylark, *Alauda arvensis.* In Ireland the Hooded Crow greatly predominates and is known to breed, possibly or probably in all but six of the 10 km squares. Breeding of the Carrion Crow in Ireland has only been confirmed in four of the 10km squares and occurs only sporadically in a few others. The Hooded Crow is found generally in open countryside of all types but less often in open types of woodland. It needs no description and the grey areas are as indicated in the illustration.

It usually nests high up in a tree but in the absence of a tree, it can be on a sea-cliff or a steep bank. Four to six eggs are laid at the end of April and have a light blue to green ground colour, blotched and spotted with umber-brown and with ashy shell marks. Incubation takes 19 days and the young are fed mostly with re-gurgitated food, being fledged in from four to five weeks.

The Hooded Crow and Carrion Crow also possess an unenviable reputation as far as their food is concerned. It is true that they consume a certain amount of carrion of great diversity, but they will also attack and kill wounded birds which they come across, and a number of healthy birds are recorded as suffering the same fate. Perhaps they are most disliked on account of their predominant egg diet in spring and early summer. They are also reputed to kill lambs. They certainly kill weaklings, and they eat the afterbirth.

Rook *Corvus frugilegus*

The Rook is a well-known bird almost through-out the whole of Ireland. It is a commensal of man, appearing to be closely bound up with his proximity or, at least, with his agriculture. It is a colonial nester — almost always nesting high in trees, often beech, *Fagus sylvatica,* and Scots pine, *Pinus sylvestris,* on the outskirts of, or even in the middle of villages or small towns. In Ireland, the number of nests constituting a colony is relatively small (there are not many colonies larger than 50 pairs), but the colonies are thought to be more numerous than in Britain. The tendency appears to be towards smaller colonies generally.

Many people have difficulty in differentiating

left: Rook, *Corvus frugilegus.* Total length 45 cm. Widespread resident increasing in abundance and breeding in every suitable locality.

right: Chough, *Pyrrhocorax pyrrhocorax.* Total length 27.5 cm of which the beak may be up to 5 cm. Precipitous cliffs and rocky islands, generally in the South and West, are the home of this resident. Also breeds inland in a few places in Counties Galway, Clare, Donegal and Kerry. Now recovering after period of decline.

below: Jackdaw, *Corvus monedula.* Total length about 32 cm. An abundant resident breeding in every county. Increasing in numbers.

between the Rook and the Jackdaw, *Corvus monedula,* indeed both are usually known as crows. The greyish-white, bare face of the Rook can be seen at considerable distances, contrasting quite strongly with the black face but grey neck of the jackdaw. This colouration makes the Rook's beak appear longer than it is in reality. When on the ground, the 'baggy trousers' effect of the loose feathers on the flanks of the Rook are characteristic. Then, of course, a Rook 'caws' whilst a Jackdaw 'tchaks'.

Jackdaw *Corvus monedula*

The Jackdaw, whose noisy 'tchaking' groups often give some movement to our ruined buildings, church towers and craggy cliffs, is a widely distributed bird in Ireland. There are few localities where it is not known as a breeding bird. Indeed, in some districts the Jackdaw is a most undesriable member of the wild fauna. This is when it builds its nest in a chimney. Often a prodigious amount of twigs and debris is dropped into the flue in order to bring the top of the nest fairly near to the entrance hole. A chimney not in use in early spring rapidly becomes unusable, so that metal mesh or, preferably, iron bars are fitted to chimney pots in order to discourage the nesting Jackdaws. Due to the acid flue-gases, corrosion takes place rapidly, so that wire mesh scarcely lasts a season.

Four to six eggs are laid from mid-April, and only the hen incubates whilst the cock feeds her on the nest. The young hatch after 17-18 days and are thereafter fed for 30-35 days on food regurgitated from a throat pouch.

The identity of the Jackdaw is unmistakable if a glimpse is taken of its dark grey neck and ear-coverts, and its rather greyish underparts with the nape, crown and wings a sooty-black. Its pearl-grey or bluish-grey eye is most marked and its beak is smaller than that of most CORVIDAE with which it is likely to be confused.

A point of great interest concerning the Jackdaw is its habit of hiding objects — not necessarily food — when it lives in association with buildings. The Jackdaw of Rheims will be familiar to many readers with its taking and hiding a ring. The old, poetically related ballad is, no doubt, founded on a basis of fact.

The Chough *Pyrrhocorax pyrrhocorax*

The Chough is a member of the family CORVIDAE and the subfamily CORVINAE which contains the largest of the perching birds — the Raven, Crow, Rook and Jackdaw. Its beak is strong and powerful but appears more delicate than those of the other members of that family. In the CORVIDAE the first primary feather is always much shorter than the second. The nostrils are always covered with forwards-directed bristles. The young closely resemble the adults.

Whilst in the subfamily CORVINAE the plumage is black and sombre, in the other subfamily GARRULINAE (containing the Jay), it is much more colourful. Members of both groups, however, are noisy and assertive and are non-specialized, being able to eat almost anything. They have a wide distribution and their aggressive, 'cheeky' behaviour places them amongst the most well-known of all birds. Within the British Isles, however, the recent history of the species with which we are concerned here, the Chough, shows, alas, a great diminution of its range.

Within the last few years it has disappeared from the cliffs of Cornwall and is now no longer an English breeding bird. Today the Chough occurs in the mountainous regions of Snowdonia, where it nests in old quarries and disused mine shafts. It is found also on some islands off the Welsh coast, and a few colonies survive in isolated cliff-side areas in southwest Scotland. It is in Ireland that the Chough still remains a dominant bird of the wild cliffs, both on the sea shores and in a few inland areas.

It is an elegant bird, 35-37cm in length, and much sleeker in appearance than any other member of the CORVINAE. All its plumage is black but is irridescent with blue, purple and green. The long, curved, sharply-pointed beak with forceps-like tip, and the feet, are vivid vermilion.

Perhaps it is in flight, that barely brooks description, that the Chough is recognized; mostly in hair-raising situations for human beings, where hand and foot-holds are scarce. The birds glide into view (usually in pairs), whence they proceed to dive maybe for a hundred metres before flattening out and floating like black feathers, seemingly only a few centimetres above the white-capped waves smashing onto the rocks below. Even the most appalling weather makes little difference to their display.

In flight, the Chough has a compact appearance. Although the tips of the primary feathers, generally splayed out, give a sharp appearance to the wings, the long secondaries make the base of the wings broad, giving this compact shape. The tail is 'squared-off'.

Nesting takes place on the ledges on cliffs, and in caverns and similar situations which are generally inaccessible. Old buildings are sometimes utilized. The nest is a bulky mass of heather, furze, bracken and other dead plant material. It is lined with sheep's wool or hair. The eggs are variable in number from two to six, but usually three or four are laid. They are white, yellowish or pale green in ground colour but have ashy blotches and are mottled and spotted with sepia and yellowish-brown. The hen incubates from the time of laying the first egg, and she is fed by the cock. Hatching takes place after 17 to 18 days, and the young are fed by regurgitation of a wide variety of insects of the soil, although ants and their larvae are said to be of great importance in the diet. The adults carefully clean their beaks and plumage after each regurgitation.

With a description of its cry, let us conclude this account of the Chough. The bisyllabic *Chee-ow* can often be heard when the birds are out of sight. It is pitched high and reverberates around the cliffs, so that it is easily heard and distinguished from the monosyllabic *Tchak* of the Jackdaw, and is so indicative of the wild places it inhabits.

Wren *Troglodytes troglodytes*

The Wren is the third most widespread bird in Britain and Ireland today. It is exceeded only by the Lark, *Alauda arvensis,* and the Hooded/Carrion Crow, *Corvus corone.* It is also the most common nesting bird, as shown by the recent Nesting Bird Census. It is extremely adaptable, being found in almost every type of habitat — from rocky islands to mountains and moorland. Although the Wren population is severely hit by long periods of cold weather it has an extra-

Wren, *Troglodytes troglodytes*. Total length just over 9 cm. One of the most generally distributed residents breeding in every county.

ordinary capacity for recovery. In deciduous woodland there may be as many as 100 pairs per square kilometre, and the total population in Britain and Ireland has been put at 10 million pairs!

The Wren is a busy, assertive little bird as it forages for insects and spiders on or near the ground, making its way mouse-like through scrub and ground litter. It seeks shelter for roosting amongst ivy, in holes in trees, haystacks or thatch, in hen-houses, nestboxes and old nests. It is not generally gregarious, but up to a dozen or so roost together in a small area on occasion. The domed nest is beautifully constructed, several being built by the cock bird, with the hen lining the chosen nest with feathers for the breeding.

The Irish Dipper
Cinclus cinclus hibernicus

In general appearance Dippers are Wren-like, with a short, stumpy body and upturned, stubby-like tail. They are larger in size, being approximately 17cm in length. They possess stout legs. Dippers are the only truly aquatic perching songbirds and show a number of interesting adaptations to the highly specialized life around fast-running and often torrential streams in mountainous districts.

The soft, filmy plumage is supported by a thick, downy undercoat, and an extremely large preen gland (which secretes an oily substance) is present for waterproofing the outer feathers. Their habit is to swim underwater, using the wings for propulsion, and to walk along the stream bed, always upstream in order to neutralize their buoyancy. They remain submerged for a maximum period of thirty seconds and more usually about ten seconds. A flap over the nostrils can be closed in order to keep out water. Their 'third' eyelid, or nictitating membrane, which is brought down over the eye to clear away water after searching the stream bed for food is often conspicuous. Their food consists of aquatic insects and their larvae, crustaceans and sometimes small fish.

The flight of the Dipper is fast and straight, usually being confined to the course of the stream and only a metre or so above it. It is a solitary bird except in the breeding season. It has a rather sweet song which it utters usually from a stone in the middle of the stream, or when in flight it makes a clinking sound rather like two pebbles being knocked together. A wren-like habit is the building of a domed nest with a side entrance. The nest is always close to the water and generally immediately above it. Although crevices in rocks and amongst tree roots are the usual locations, the dipper often utilizes man-made situations such as the underneath of a bridge.

The Dipper has a chocolate brown head and nape. The rest of the body (with the exception of the throat and upper chest, which are white), is greyish brown with a greyish edging to the feathers, thus giving a scaly appearance. However, the British race, *Cinclus cinclus gularis,* shows an unmistakable chestnut area on the edge of the white 'bib', which is absent in the Irish sub-species. In addition, the general body colour of *C.c. hibernicus,* including the crown and nape, is much darker in colour.

The distribution of the Irish dipper is interesting in that it covers not only the whole of Ireland but also the Isle of Man, as well as the west coast of Scotland, the Outer Hebrides (Harris, Lewis and Barra) and the isles of Bute and Arran.

Stonechat *Saxicola torquata*
This little bird is characteristic of gorse-covered hillsides and other uncultivated areas, and is widely distributed throughout the country, although generally local. In some areas, where

gorse is absent, it is a hedgerow bird, but it is also found on commons, heaths and in young forestry plantations, wherever there is heather. Except for an inland area in the north-east around Lough Neagh and for some scattered localities in the south-east, it is known as a breeding bird throughout Ireland, with perhaps populations particularly dense near the coast.

The male perches conspicuously most often on gorse, within a short distance of a human intruder, with much tail-flirting and wing-flicking and making short, hovering flights. It keeps up its well-known 'tsak-tsak', like two pebbles clicking together. In appearance the male is most distinctive, with a black head, white patches on sides of neck and wings and with reddish-chestnut breast. The female is less distinctive. The beak, legs and feet are black. It breeds early in the year, its nest being constructed in late March or early April, and situated usually on the ground and often at the base of a gorse bush. Five or six eggs are laid, greenish-blue in colour, with rusty specks. Two or three broods are generally the rule, but as many as four are not altogether unusual.

Insects are the chief food, often quite large species such as butterflies, large moths and dragonflies being taken. Hard winters reduce the population considerably and there is a certain degree of migration.

Pied Wagtail *Motacilla alba*
This is one of the most abundant species of breeding birds of Ireland. It is widely distributed, and most country-folk are able to identify it, although it may be confused with the Grey Wagtail, *Motacilla cinerea,* but the entirely black and white colouration, its bounding flights and its quick runs and the flicking tail, make it unmistakable. It is sometimes called the Water Wagtail, since it is often found in association with streams, other aquatic and seaside environments. However, it also occurs far from water, especially where there are stone walls and in areas where fields are bounded by walls. Roads make good feeding grounds for this species. People often call the Pied Wagtail the "Willy Wagtail" while the Grey Wagtail *Motacilla cinerea,* is mistakenly called the "Yellow

Wagtail" (because it has a yellow breast). The true Yellow Wagtail, *Motacilla flava* (which is a rare species in Ireland), has an olive-green back, whereas in the Grey Wagtail it is grey.

It is generally thought that the Pied Wagtail has steadily increased in numbers during the past fifty years or so, and has extended its range into Mayo and Galway where formerly it was unknown. Now it has been recorded in every 10km square in the country, and in almost every one as a breeding bird.

Many city dwellers also know the Pied Wagtail because of its strange habit of roosting gregariously from autumn through the winter to early spring. The roosts selected may be isolated trees, reed beds, rhododendron bushes, gorse, ivy-covered walls and the roofs of buildings. Numbers roosting in these communal roosts often run into many hundreds. Perhaps the largest roost in Britain and Ireland occurs in O'Connell Street (in Plane trees) in the centre of Dublin's fair city. To see, I would estimate, about a thousand Pied Wagtails gathering in the evening gloom, is an amazing experience for anyone with an interest in nature. It nests in holes in walls or in steep banks, pollard willows or in the thqtch of buildings. Often it will use the 'open' type nest box. (This is a nest box with a wide, open slit instead of a small hole.) The five or six eggs are greyish- or bluish-white and are usually spotted with pale, lead-brown and grey. There are two or three broods annually. Incubation takes 13 to 14 days, and the young fledge within 14 or 15 days.

above: Stonechat, *Saxicola torquata*. Total length about 12.5 cm. Widely distributed resident and Summer visitor which is usually more plentiful in coastal areas.

above: Pied Wagtail, *Motacilla alba*. Total length about 17.5 cm. One of the commonest and most widespread resident breeding birds in Ireland.

left: Irish Dipper, *Cinclus cinclus hibernicus*. Total length just over 17 cm. A resident which breeds in every county.

4 Reptiles and Amphibians

Lizard, Frog, Toad, Newt

Reptiles *Reptilia*

Reptiles are an advance along the evolutionary path from the amphibians and are readily differentiated by the skin being covered with scales; the presence of single teeth in the jaws; internal fertilization; the eggs containing yolk and being enveloped by a protective membrane or a shell, as well as a number of other important anatomical features.

The eggs' protective covering enabled the reptile to lay them on dry land, so that it was not essential for the adults to return to the water for reproduction. Breathing through the skin was lost by the development of horny scales and so lung respiration needed to be more efficient and this resulted in the ventilation of the lungs by movements of the rib cage. The external development of a larval form of embryo (tadpole) was dispensed with, so that on hatching, the egg produced a hatchling resembling its parents in almost everything except sexual maturity and size.

Between 300 million and 70 million years ago, reptiles were the dominant animals on the earth and many grew to enormous size. During the last 70 million years, however, reptiles have been in decline, although there are about 6,000 species still living and of these about 2,500 are lizards. Ireland possesses one species only — the Viviparous lizard, *Lacerta vivipara*.

Viviparous Lizard *Lacerta vivipara*

This attractive, brown-patterned lizard is Ireland's only reptile and is widely distributed throughout the country, although it appears to be less abundant in the inland counties. Occuring on Cape Clear Island, it may well be present on other islands and is worth looking for and recording, if found. It occurs throughout Europe's Central and Northern areas — even inside the Arctic Circle — and it is said to show a preference for mountainous country.

It is a small lizard, probably averaging less than 10cm in length. The females are larger and much stouter generally, with the abdomen exceptionally so, and the tail is much less graceful than that of the male. The male can also be distinguished from the female by the presence of a swelling at the base of the tail. In both sexes the body is covered with scales which are smaller on the back and much larger underneath, and they hardly overlap. The scales on the top of the head are large also and the eyes are furnished with eyelids. The tongue is notched and ends in two rounded lobes. The body is variously marked with longitudinal stripes in shades of brown, but there is usually a dark band running down the back which has a central row of lighter markings and similar lateral stripes. The tail possesses a built-in safety device in that a piece of it will snap with ease if seized by a predator – and this includes man – enabling the lizard which has been caught unawares to make an unexpected escape! Some regrowth occurs.

The Viviparous lizard mates in April and May without any preliminary courtship, and after approximately three months gestation the young lizards are born. The brood of from four to ten young, 4cm long, may take some days for all to see the light. The young lizard is able to fend for itself within a few minutes of birth and is able to eat very small insects such as aphids. As it grows it is capable of tackling larger and larger prey. This lizard is a delightful little animal to watch. When in areas where it is known to occur, keep very still, wait and keep watch for it to appear.

AMPHIBIANS *Amphibia*

Frog, Toad and Newt
Amphibians were the first back-boned animals (vertebrates) to emerge from the water onto dry land. This was about 350 million years ago in the Devonian period. However, they have never lost their dependence on an aquatic environment. Firstly, the skin is thin and not protected with scales against desiccation, so that the immediate environment must be wet or at least damp, with water close at hand. The skin contains glands which secrete mucilage, and this helps to keep the skin moist. Often poison glands are present and are particularly prevalent in toads. Part of the life-cycle is generally spent in water or in conditions which approximate to it. Secondly, the eggs are naked, there being no membrane or shell to resist drying up, so that they must be

Frog, *Rana temporaria.* Length about 10 cm. Widely distributed throughout the country.

laid in water or at least in humid conditions. The larval state of amphibians (the tadpole), is very fishlike with the sensory organs in a lateral line down the sides of the body, just as in fishes, and in the possession of gills for respiration and a tail-fin. Indeed, in many species, the tail-fin is maintained throughout the life of the amphibian.

The adult breathes both by means of lungs and through the skin. The air is forced into the lungs by the movement of the base of the mouth cavity, which is activated by muscles, and not by rib-movement as in the case of reptiles, birds and mammals.

Frog *Rana temporaria*
Whether the Common frog, which is so widely distributed throughout Ireland, is an introduction by man, or whether it is a native, is a point over which controversy still requires the experts to give their views. Professor O'Rourke believes that the evidence is largely in favour of

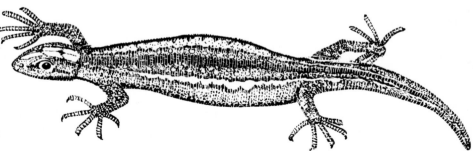

Viviparous Lizard, *Lacerta vivipara.* Average length about 12 cm but female may attain length of about 17 cm. Widely distributed and abundant throughout the country.

the view that it was introduced into Ireland about the time of the Normans. After nearly 900 years, surely they can be considered Irish! However, it appears certain that Dublin had no frogs until a Fellow of Trinity College brought some frog-spawn and planted it in a ditch in the grounds of the College.

The general appearance of the Common frog is familiar to almost everyone, although the ground colour is very variable. This is on account of the frog's ability to vary the basic colour of its skin according to the nature of its immediate environment. Then, the cells of the skin expand or contract causing the pigment to vary with the intensity of the reflected light. Its forelegs are short and the four digits are not webbed, contrasting with the very long hind legs with five webbed digits. These enable the frog, not only to be an excellent swimmer, but also a good jumper. The skin is maintained in a moist condition by the secretion of innumerable mucous glands with which it is abundantly supplied. Oxygen is absorbed through the skin as well as by the lungs, but air must be forced into the latter by the action of swallowing.

The process of reproduction takes place early in the year, usually in February and March. This is preceded by a migration to the breeding ponds, and when several hundred frogs are to be seen crossing a road, it is quite an interesting event. The frogs seem to have some direction-finding sense which aids them in finding a suitable location for depositing the mass of spawn – the eggs – which have been externally fertilized by the male as he grips the female in a vice-like hold around her abdomen, ejecting sperm over the eggs as they leave the female's urino-genital aperture. Between 1,000 and 4,000 eggs are spawned, and immediately they are deposited, their gelatinous covering swells up on absorbing water, giving the familiar appearance of 'frog-spawn'. In about one month, the black egg in the centre will have developed into an embryonic tadpole which clings to the outside of the jelly by means of suckers. The subsequent development into a wriggling tadpole; the emergence of limbs and the reduction in the tail to give a minute frog, are so familiar that the details will not be given here.

The food of the frog consists of many kinds of insects, slugs, worms and other invertebrates which are caught by the frog shooting out its sticky tongue from the floor of the mouth with great rapidity, and flicking it back in again with the prey adhering to it. In Britain, the frog has suffered a great decline in numbers, and what a great pity it would be if a similar fate befell it in Ireland.

Kerry Toad *Bufo calamitas*

This small toad is confined in Ireland to certain coastal areas in Kerry and it may exist in Co. Sligo. In Britain, where it is called the Natterjack, it has largely disappeared from many of its old haunts and the small, remaining colonies are given special protection. Especial care must be taken of the Irish colonies of this interesting and harmless amphibian.

The Kerry toad grows to about 7cm in length and its basic ground colour is a pale, yellowish-brown or dark green. It is spotted and blotched with a darker brown or dark green, and underneath it is a very pale yellow or yellowish white with black spots. The legs are also barred with black. The conspicuous eyes are greenish-yellow

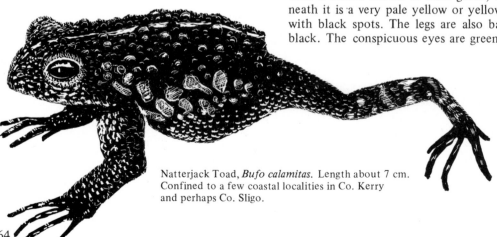

Natterjack Toad, *Bufo calamitas*. Length about 7 cm. Confined to a few coastal localities in Co. Kerry and perhaps Co. Sligo.

Common Newt, *Triturus vulgaris.* Male attains a length of up to 10 cm. Thought to occur throughout the country.

element in the Lusitanian fauna. But doubts have been expressed recently that this is so and a suggestion has been made that it was introduced at some time.

but the most prominent feature is a narrow yellow line which runs along the middle line of its head and back. Its method of progression is characteristic since the hind legs are not unduly long so that it is unable to jump, moving in a series of short runs punctuated by pauses. Its skin is generally very warty and wet-looking due to copious secretions of mucilage — quite unlike the Common toad, *Bufo bufo,* of Britain which is rough in appearance but dry-looking. The Kerry toad is said to be darker than the British natterjack.

Found in marshy, generally sandy areas, which frequently occur behind coastal dunes, the Kerry toad burrows into the damp sand and usually spends the daylight hours just beneath the surface. Occasionally it may run over the hot surface in full sunlight, seeming to be able to withstand such conditions much better than its counterpart, the Common European toad. In spite of its short back legs it is a good swimmer.

The Kerry toad appears to be somewhat gregarious, as when one or two are seen, and many more may be found in the same locality. When alarmed it can secrete a whitish fluid of a characteristic odour and will even sham death by stretching and flattening itself out on the ground. The males produce a loud, rattling croak at breeding time which can be heard over a long distance, from the end of April to June. The 'spawn' — the string of eggs — may be 2m in length, and when fully grown, the tadpoles metamorphose into 12mm toads in less than six weeks. Growth is slow from then onwards, and the toads mature in four or five years and average between 38mm and 50mm in length.

Praeger, the well known naturalist believed that the Kerry toad was a relict species — an

Common or Smooth Newt
Triturus vulgaris

Although the precise distribution of the newt in Ireland is not known, it is thought to occur throughout the country. Reaching a length of up to 10cm, it is variable in colour. The ground colour is olive-brown or greyish-brown with series of dark spots on the upper-side. Underneath, the newt is silvery-grey suffused with orange or vermilion, and rows of circular black markings are usually present. The colour is much more intense in the breeding season — spring and early summer. Early in the year, the males develop a crest along the back and the tail, whilst a blue band appears along the lower edge of the tail. The top of the crest has a wavy edge. The male has longer digits than the female.

In early spring, both sexes seek water, and during the whole of the breeding season are to be found only in ponds containing water-plants. Periodically they must come to the surface for air. An elaborate courtship takes place, after which the male deposits a mass of sperms (the spermatophore), which sinks to the bottom of the pond and the female takes it into her body through the cloaca, the common urino-genital aperture. On laying the fertilized eggs, the female wraps each one in the leaf of a pond-weed. They hatch in about two weeks into rather slender tadpoles which are furnished with gills and a pair of organs arising from the upper jaw which enable the tadpole to cling to the water plants. Development is completed by the end of the summer when they leave the water and hibernate under stones. Newts feed on a wide variety of insects and other invertebrates. Anything which moves near it is eyed with interest as a probable meal! The newt was the subject of much myth and superstitition in days gone by.

5 Fishes

Fish are cold-blooded, vertebrate animals furnished with fins, breathing by means of gills, and confined to an aquatic environment. World-wide there are about 20,000 different fish species, so that there is a great variety of shape and habit, although all are generally recognizable as fishes. The shape of the body is streamlined such that little resistance is offered to it when swimming. With few exceptions (one being the Conger eel), the body is covered with over-lapping scales. These are small, bony segments fixed into the skin in a regular pattern, and the age of a fish can often be determined by counting the concentric circles (growth rings) formed on the scales. In the case of sharks and rays, however, the scales do not overlap. They are teeth-like and sharp, giving the skin an extremely rough feel.

The fins of fishes consist of folds of skin supported by a series of stiff spines or softer rays, and usually they can be erected at will by muscles and articulations at the base of each supporting spine. The numbers and positions of the fins follow a general pattern, although one or more may be absent. There is a pair of fins — the pectorals — situated behind the head, one on each side, and similarly a pair of pelvic fins, but their relative positions may, however, be reversed. A series of dorsal fins is located along the mid-dorsal lines and, likewise, anal fins occupy the mid-ventral line. The caudal fin is vertical (unlike that of whales and dolphins), and composed of dorsal and ventral elements.

Forward propulsion is brought about by alternate side-to-side movements of the body in which the caudal fin takes some part — the other fins functioning as stabilizers, brakes or rudders except in some special adaptations. Most fishes possess a 'lateral line' consisting of a series of sense organs (usually visible as a stripe) along the sides of the body. It is often of importance in identification.

The gills of fishes consist of a large number of thin-walled folds of skin in which there is a rich blood supply and these are supported on cartilaginous arches. When water is taken in at the mouth it passes over the gills where oxygen diffuses in and is absorbed by the blood stream whilst carbon dioxide is removed. In other than sharks and rays, the gills are covered by a bony plate which protects them. In sharks and rays the gills, in the form of slits, do not share a common protective shield.

The Basking Shark
Cetorhinus maximus
The Basking shark or 'muldoan' as it is called in Ireland, is one of the largest fishes to be found today roaming the oceans. Indeed, its 9m (30ft)

Basking Shark, *Cetorhinus maximus.*
Length about 9 m. Often common when
on migration in western and southern
coastal waters.

below: Basking Shark, *Cetorhinus maximus.* It is known occasionally to leap clear of the water.
This has been observed by the Author, and it re-entered the water with a sound like a rifle shot.

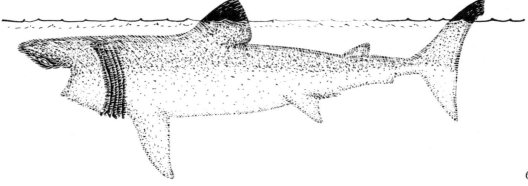

places it second only in size to the Whale shark, *Rhinodon typus,* which is known to reach a length of 13.5m (45ft) and specimens of even greater length have been reported. But, whereas the latter is confined to the warm waters of the world, the Basking shark is most often found in the temperate seas of the northern hemisphere. Most importantly for us, these great fishes include the west coast of Ireland in their itinerary.

So unlike are they in the essential characteristics, these huge sharks are placed in separate families, but their feeding habits are identical in that they feed on planktonic fauna. These are sieved out of the water by the gills which are equipped with a series of fine rakers, up to 10cm in length, of which over a thousand are present in each row. All the water coming into the mouth and leaving by the gill clefts is strained of all its organisms — mostly crustaceans. Sometimes Basking sharks are found without gill-rakers and it has been presumed from this that the rakers are shed and replaced periodically.

The Basking shark is so named from its habit of resting almost motionless at the surface with its large front dorsal fin visible above the surface of the water. It is greyish-brown in colour with a rather protuberant snout, but the most characteristic feature is the enormous size of the gill clefts which stretch from the ventral to the dorsal surfaces of the body. It is not aggressive, being harmless to man, unless its sheer size causes it to accidentally upset small boats which may venture too close. Its teeth are small. A weight of 4,000kg (4 tons) may be reached. It appears to be generally distributed in the North Atlantic, migrating along the western coasts of Ireland and Scotland, as well as along the New England coasts of the United States of America. In the Pacific it moves along the coasts of South

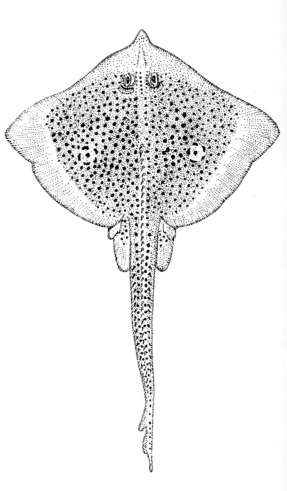

above: Thornback Ray, *Raja clavata.* Length of male about 70 cm but of the female 120 cm. Common throughout Irish waters.

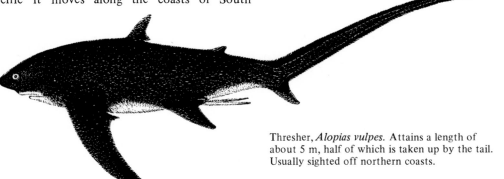

Thresher, *Alopias vulpes.* Attains a length of about 5 m, half of which is taken up by the tail. Usually sighted off northern coasts.

America (Peru and Ecuador) and California. There are comparatively few records of its sighting in Asiatic waters.

The Basking shark is usually caught by harpoon as it basks on the surface, and is fished for its liver, which weighs about a tenth of its total weight. The liver-oils are important in leather tanning processes. Its breeding habits are virtually unknown, but it is believed to be viviparous. The Whale shark, on the other hand, lays eggs, the egg-cases being about 60cm (2ft) in length.

Thresher *Alopias vulpinus*

The sharks and rays all belong to one of the major groups of fishes, the *Selachii.* Their main characteristic being that the skeleton is of cartilage, whereas in all other fishes it is of bone. There are five, but on occasion six, gill slits. In this group, Ireland possesses a substantial number of species which have become of importance in rod and line angling. In 1932 a Porbeagle, *Lamna nasus,* weighing 65kg (365 lbs.), was taken at Achill, and a Blue Shark, *Prionace glauca,* of 93kg (296 lbs.), was caught in the same area in 1959. A number of other 'deep-sea' fishing locations have been, or still are being developed, especially on the south and west coasts of Ireland.

The Thresher is identified by the great length of the upper lobe of the tail fin. The snout is relatively short, and the un-serrated teeth are triangular, whereas they are long and spear-shaped in the Porbeagle — the only species with which the Thresher could be remotely confused, at least if caught by angling. The maximum length of the Thresher is about 5m (one being taken off Donegal in 1905), and about half of this is taken up by the tail.

It is, however, harmless to man and appears to feed mainly on herring and mackerel, and on occasion the Thresher has been caught in nets put out for these two species. Most sightings of this fish have been off the northern coasts, and they are known to swim around 'shoaling' fish, when the Thresher will 'thresh' the water with its tail in order to concentrate its prey before moving in to feed on the shoal. One record is of 27 mackerel being found in the stomach of a Thresher!

Thornback Ray *Raja clavata*

Eleven species of rays and skates (family RAJIDAE) are recorded from Irish waters. In addition, the Electric ray, *Torpedo nobiliana* (family TORPEDINIDAE), is sometimes recorded from the southern and western coasts, and the Sting ray, *Dasyatis pastinaca,* (family DASYATIDAE) off south and southwestern coasts. They are related to the sharks, in that the skeleton is of cartilage, but they differ from them in that the body is flattened and the pectoral fins are large and reach the snout, giving the body a rhomboidal shape. The hinder end is long and tail-like. In the Thornback ray, the upperside is spotted and a number of thorn-like spines are present, especially along the mid-dorsal line. In the specimens which the author has examined in southwestern Cork, there is an eye-like mark on each side of the flattened body, although there is much variation in its distinctiveness.

The male attains a length of about 70cm and the female 120cm. Usually found over mud at depths of about 20 to 100m, they feed on animals such as crabs, prawns, small flatfishes of varied species and on Sand-eels, which it can literally 'drop' upon! It is not until the male is about 60cm in length that it is sexually mature, and is easily identified by a pair of rod-like parts, serving as intromittent organs or penes (plural of penis).

During the course of a single summer, the female lays only twenty eggs which are heavily protected from predation. The egg capsule is rectangular and flattened, and the corners are extended into a hollow, horn-like point at each corner, and the capsule is about 6x4cm in size. Oxygen for respiration is taken from fresh sea-water which enters through slits in the capsule wall. The eggs take up to about five months to hatch. Capsules of this type are produced by all the rays and skates of the family RAJIDAE, and are different from the egg-capsules of some dogfishes, with the latter having long, coiled tendrils at the corners. The Thornback ray illustrated was the first animal drawn and then eaten by the author afterwards!

Sturgeon *Acipenser sturio*

From the classification point of view, the Sturgeon is in an isolated position. It is exceptionally primitive amongst the 'bony' fishes — those fishes other than sharks and rays — and is little changed since Palaeozoic and Mesozoic times. There are 26 different species of Sturgeon in a single family, the ACIPENSERIDAE, the latter being the sole family in the order *Chondrostei*. Most Sturgeon species occur in Russia and Asia, only one species being found in Western Europe. It is now considered scarce in Irish waters, although formerly it was more common, but was probably never a very abundant species in Ireland.

Most of its life is spent in the sea, but it ascends into freshwater up the larger rivers in spring and early summer, at the time when it is sexually mature (from seven to 14 years old) when the eggs are laid and fertilized. This is the time when the fish are caught in Russian rivers — principally the Volga, when roes are used for the delicacy 'Caviare', and the flesh is highly prized. The Sturgeon grows to a large size and although 1 to 2m is now the average size, specimens are known of 5 to 6m, and its length of life is over 100 years!

In general appearance it is shark-like, but the head and body are covered with dermal bones. The upper lobe of the tail is also shark-like, in that it is much the larger. The front edge of the pectoral fins consists of a strong spine. The mouth is tubular and can be protruded for sucking up small animal life, which it digs up with its sharp snout, sensed by a group of four barbels situated between the snout and mouth.

Killarney Shad

Alosa fallax killarnensis

The family CLUPEIDAE contains the herring, *Clupea harengus;* sprat, *sprattus sprattus;* and the pilchard, *Sardina pilchardus* — each of great fishery importance. In addition, the anchovy, *Engraulis encrasicolus,* is found, from time to time, in Irish waters. All of the above are exclusively marine, but the shads, the Allis shad, *Alosa alosa,* and the Twaite shad, *Alosa fallax,* additionally spend some time in freshwater. In the case of the former species there are few records of it being found in freshwater, but it does occur occasionally in estuaries. On the other hand, the Twaite shad, is common in the estuaries of the Rivers Slaney, Barrow, Nore, Suir, and Cork Blackwater.

The Killarney shad consists of landlocked populations of the Twaite shad found only in the Killarney Lakes and are sufficiently distinct from those able to migrate from sea to freshwater to merit sub-specific rank. Both shad species are herring-like in form, but the Twaite has six to ten black spots on each side, and the first gill-arch is furnished with 40-60 gill-rakers. The Allis shad has only one to six, more or less, distinct spots on each side, and the first gill-arch bears from 90 to 120 gill-rakers. The main distribution of the Twaite appears to be off northern Germany, but the Allis is most abundant in the Bay of Biscay and in the Mediterranean.

Char *Salvelinus alpinus*

The char, is a member of the SALMONIDAE, the salmon and trout family, and occurs in a number of cold, deep loughs. Indeed, it was found in even greater numbers until comparatively recently. It seems certain that formerly the char was migratory, rather like the salmon which ascended rivers from the sea and spawned in small streams — often at considerable altitudes. This habit is still shown by Arctic char species. It is thought that present day Irish char are derived from landlocked populations, and it is probable that they still seek out the inflowing streams for spawning. The isolated populations have given rise to a number of variations, each of which, at one time, was given specific rank — but, today only a single species is recognized.

It is a most variable species as far as colour is concerned. The back has been described as bluish-black, or olive, green or brownish. The belly is silvery-white or orange or crimson, and in the males, in the spawning season, a deep red colour is often shown. The presence of spots of pink, orange, or red is general, but their numbers may be reduced or be absent altogether.

Cole's char, *ssp.colii,* occurs in Loughs Eask

in Co. Donegal, Dearg in Connaught, and in Lough Currane in Kerry. In the larger loughs it may reach a length of 30cm. Gray's char, *ssp. gravi,* occurs in Lough Melvin in Co. Fermanagh, where it was known as the Freshwater herring. Travelyan's char, *ssp.travelyani,* from Lough Finn in Donegal; Scharff's char, *ssp.scarffi,* from Lough Owel in Westmeath; the Coomasaharn char, *ssp.fimbriatus,* from Lough Coomasaharn in Co. Kerry, and the Blunt-nosed char, *ssp.obtusus,* from Loughs Luggala and Dan in Co. Wicklow and Killarney, and Accoose in Co. Kerry, may all be differentiated from each other by their varying forms and colours.

Pike *Esox lucius*

The pike family, ESOCIDAE, occupies a more or less isolated position in the fish classification, although it is usually placed between the SALMONIDAE and the carp family, CYPRINIDAE. The pike is a common and very important anglers' fish occurring in freshwater — both rivers and lakes. The elongate body is covered with small scales. Barbels and adipose fin are absent, and the dorsal fin is located far back, opposite the anal fin. The head is snout-like and the gape of the mouth is large. The strong teeth, at the edges of the lower jaw, are well-suited for seizing prey. In colour it is greenish on the back and sides, with the under-parts white. A distinguishing feature is that both back and sides are liberally spotted or banded with yellow. The dorsal, tail and anal fins are spotted and striped. It is the largest freshwater fish occurring in Ireland, often attaining a weight of over 11kg (25 lbs.). The heaviest river Pike caught by fair angling was one of 19kg (42 lbs.) and 17kg (38 lbs.) is the record for Lake pike, although a specimen of 24kg (53 lbs.), caught in Lough Conn in 1920, is given in some accounts.

The pike is a ferocious predator, for although mainly feeding on other fish, it is known to attack and swallow water-birds, rats and other animals swimming in their haunts. Izaak Walton's famous recipe for cooking pike ends with the comment, "Too good a dish for any but anglers, or very honest men"!

Tench *Tinca tinca*

The tench is a member of the CYPRINIDAE, the carp family. Other Irish species are the gudgeon, *Gobio gobio*; the rudd, *Scardinius erythrophthalmus*; and bream, *Abramis brama*; the minnow, *Phoximus phoximus*; which, although locally common, has an expanding distribution; the dace, *Leuciscus leuciscus,* introduced into the Cork Blackwater as long ago as 1889 and now abundant. The carp, *Cyprinus carpio,* introduced into a number of lakes is now plentiful in the Lough in Cork City, and the roach, *Rutilus rutilus,* also introduced into the Cork Blackwater in 1889, is now found also in the Fairywater Erne and Shannon. Hybrids with the latter species, and the bream or rudd, are also known.

Although not usually distributed widely, the tench is common where it occurs. In colour it is very dark, covered with small, golden scales. The fins are large and oval. It is a fish which is to be found only in still water at the bottom where water-weeds are plentiful, and this is also a muddy bottom into which it can disappear during the winter months. The heaviest tench caught by fair angling, was one of just over 3.6kg (7 lb. 13 ozs.), taken in the Shannon in 1971.

Great Pipefish *Syngnathus acus*

Six species of pipefish occur around the Irish coast. They are all extremely elongated — almost eel-like — and all possess a long, tubular snout with the small mouth at the very end. The snout acts like a pipette, sucking up prey which consists of small crustaceans as well as fish fry. The male attains a length of about 46cm with the females somewhat smaller. They are usually to be found in the seaweed zone and are remarkable for the way in which the female, after laying 200-400 eggs, transfers them to the brood pouch of the male. This pouch is a fold of skin situated ventrally and opposite the commencement of the dorsal fin. The eggs are 2.4mm in diameter and after about 5 weeks they hatch into young about 30mm in length.

below: Sturgeon, *Acipenser sturio*. Usually between 1 and 2 m but specimens are known of immense size. Now scarce in Irish waters.

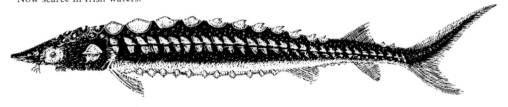

below: Killarney Shad, *Alosa fallax killarnensis*. Length up to about 40 cm. Found only in Killarney lakes.

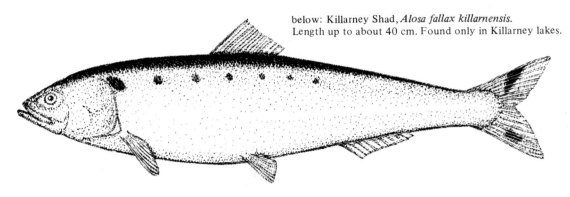

below: Char, *Salvelinus alpinus*. May reach a length of 30 cm. Widespread in a number of the larger Loughs but only occasionally caught.

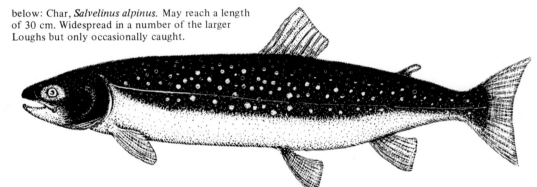

below: Pike, *Esox lucius*. Is known to attain a length of 1.3 m. A common fish occurring in lakes and in large rivers.

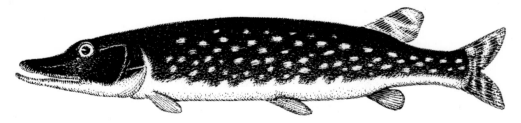

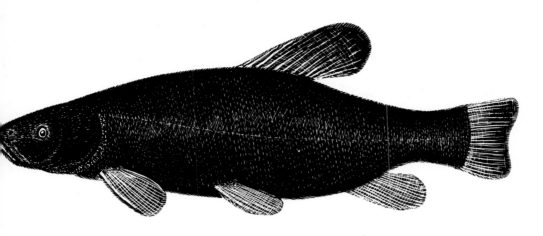

above: Tench, *Tinca tinca.* The heaviest specimen caught by fair angling weighed 3.67 kg. Common locally and generally in deep, weedy pools.

below: Great Pipefish, *Syngnathus acus.* Male up to 46 cm in length, female generally smaller. Common in seaweed beds all around the coast.

below: Lesser Sand Eel, *Ammodytes tobianus.* Reaches a length of about 16 cm, maximum 20 cm. Common on sandy bottoms from 0-30 m.

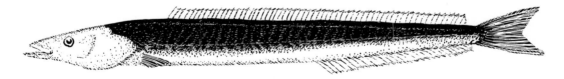

below: Ling, *Molva molva.* Length up to about 1.8 m. Common throughout Irish waters.

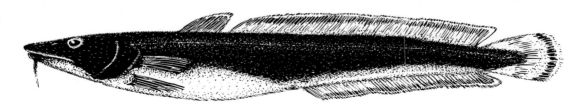

Ling *Molva molva*

The family GADIDAE, in which the ling is classified, is one of the most important economically as it also includes cod, *Gadus morhua*; haddock, *Melamogrammus aeglefinus*; whiting, *Merlangius merlangus*; hake, *Merluccius merluccius,* as well as a number of other species. Ling attains a larger size than any other member of this family — the record caught by fair angling in Irish waters being one of 21.1kg (46½ lbs.) caught off Kinsale in 1965. In length it usually reaches 1.8m and a weight of 30kg.

The Ling has an elongate body with a short, triangular dorsal fin, but with long second dorsal and anal fins. The latter is not confluent with the tail fin. The paired pectoral and pelvic fins are close together, the latter slightly to the fore. An important distinguishing feature is the length of the single, long barbel, arising from the centre of the lower lip. It is longer than the diameter of the eye.

The usual habitat is at a depth of from 100 to 600 metres, off the western and southern coasts. It feeds on a number of fish species — Dublin Bay prawns, octupuses and other marine animals. The female produces a vast number of eggs, estimated at from 20 to 60 million! They are about 1mm in diameter and are pelagic. If all the eggs of just one female reached a weight of 10kg, the resulting weight of fish would be approximately ten times the total present European catch, which stands at 50,500 tonnes per year.

John Dory *Zeus faber*

This well-flavoured fish is a common component of the trawl catch off the Irish coasts. At moderate depths it swims in small shoals near the sea bed where it feeds on small fish. The latter may be attracted to the long extensions to the membrane connecting the spines of the dorsal fin, and then caught by means of the large, protrusible mouth. The body is flattened and the head is proportionately very large. There is a conspicuous black spot in the middle of each side, ringed with creamy-gold. This is the legendary 'finger print' of St. Peter when he took a gold coin from the fish's mouth. There are several accounts of the derivation of its name. The Italian name for the fish is 'Janitore' — the Janitor — and this is the usually accepted story. The John Dory's size is that of a dinner plate and yields two triangular fillets, but, it may reach a length of 60cm, giving a weight of 8kg. There are many delightful recipes, as this fish has such a delicate flavour.

Lesser Sand-Eel *Ammodytes tobianus*

Four species of Sand-eel, in the family AMMODYTIDAE, are common around the Irish sandy shores. All are of similar appearance, having an extremely elongate shape with a long dorsal fin and elongate head which has a protrusible mouth with which small crustaceans and other small animal forms are sucked up. They occur in great abundance, and constitute an important food component for fish of economic importance — such as cod and salmon.

The Greater Sand-eel, *Hyperoplus lanceolatus*, averages about 20cm in length, but may reach almost twice this size and is of greenish colouration. The Lesser Sand-eel is often found buried in sand at low tide during the day time and averages about 15cm in length with a maximum of 20cm. The Sand-eel, *Ammodytes marinus*, apparently possesses similar habits to the Lesser, but specimens are not to be found in the National Collection. The same may be said for the Sand-eel, *Hyperoplus immaculatus*. The Smooth Sand-eel, *Gymnammodytes semisquamatus,* is identified by a waved margin on the dorsal fin, and the dark-coloured head. This species averages 15cm in length.

Rock Goby *Gobius paganallus*

Gobies, which form the family GOBIIDAE, consist of 14 species in the Irish list. Most are small, from 5 to 12cm in length only, and all are shallow water, coastal forms. Often they are very numerous. Gobies are furnished with a sucker formed from the fused ventral fins. The lateral line is either absent or much reduced, but its place is taken by a number of papilla-like sensory organs elsewhere on the body. The males of some species guard the eggs and keep a current of water flowing over them so that a continuous supply of oxygen is available to the developing eggs. Irish species of goby are marine, with the exception of the commonly occurring Estuarine goby, *Pomatoschistus microps,* which occurs in brackish water.

The Rock goby is a common Irish species but is at its northern limit around Ireland — being more generally distributed in the Mediterranean region. In this species, the first 4 rays of the dorsal fin are more or less equal in length and possess a pale edge. On the other hand, Fries's goby, *Gobius friesi,* is a north-western species and although occurring around Ireland (scarcely, in the *List of Irish Fishes*), its distribution north-eastwards to the Skakerrak may occur at depths of from 20 to 40 metres.

Dragonet *Callionymus lyra*
This must be the most brightly coloured of all the fishes around the Irish coast. The head, sides, dorsal and tail fins are spotted, streaked and striped with yellow and a vivid, almost luminescent, blue. The dorsal fins are of extraordinary size. In the male, the first ray of the first dorsal fin is as long as the rest of the body, and the second dorsal fin consists of ten equal-sized rays of large size. Viewed from the top, the dragonet is tadpole-shaped — with a very large head — the eyes situated for vision upwards. The gill opening is a hole on each side. The upper jaw can be protruded, and presumably this is the manner in which it catches its food, which is made up of crustaceans and molluscs living on the sandy beds where the dragonet itself lives. The largest males are about 30cm in length.

Tompot Blenny *Blennius gattorugine*
The blennies constitute the family BLENNIIDAE and are usually small in size but have great variety of form and colour. They are not often found at any great distance from coastal waters but the family is world-wide in distribution. The distinguishing features are, first of all, the pelvic fins which are generally in the throat region, and are made up of less than five rays, or they may be absent altogether. The body is elongate with the dorsal fin (which may be made up of three separate fins), running the whole length of the back.

The Tompot blenny is identified by its fringed tentacles arising just above the eye; the 18 rays in the anal fin and the large pectoral fins. The pelvic fins are situated just in front of the latter and are finger-like and would appear to have lost their function. There is a slight concavity towards the middle of the dorsal fin. The Tompot blenny is stated to be probably common beyond tide-marks, but is not to be found in the Irish sea. It appears to be a south-western European species — southwest Ireland and southwest England being its farthest points north. Strangely, however, five of the six specimens in the National Collection are from Co. Antrim.

Streaked Gurnard *Trigloporus lastoviza*
formerly *Trigla lineata*
There are five species in the gurnard family, the TRIGLIDAE, in the Irish list. They are characterized by the head being armoured with bony plates and spines. The lateral fin is protected by a series of spines also and these are found at the base of the dorsal fins too. The most important character, however, which most fishery people would use in identifying members of this family is the division of the pectoral fin into two parts — the first consisting of three separate rays, whilst the second part is an ordinary fin. The three rays are bent somewhat, and are used much like legs as the fish 'walks' on the sea bed.

The Grey gurnard, *Eutrigla gurnardus,* is the most abundant of the gurnards, and it has been stated that it would be quite exceptional to make a haul with the trawl without capturing a few of this species. The Streaked gurnard shows the row of spines along the lateral line and along the dorsal fin. There are a number of transverse bands which are more or less conspicuous. It is a southern species and although occasionally recorded from northern waters, the south coast appears generally to be the limit of its distribution.

Montagu's Sea Snail *Liparis montagui*
Two species only of Sea snails (family LIPARIDAE), are known from Ireland. These are the rare, but generally called Common Sea snail, *Liparis liparis,* and Montagu's Sea snail, which is

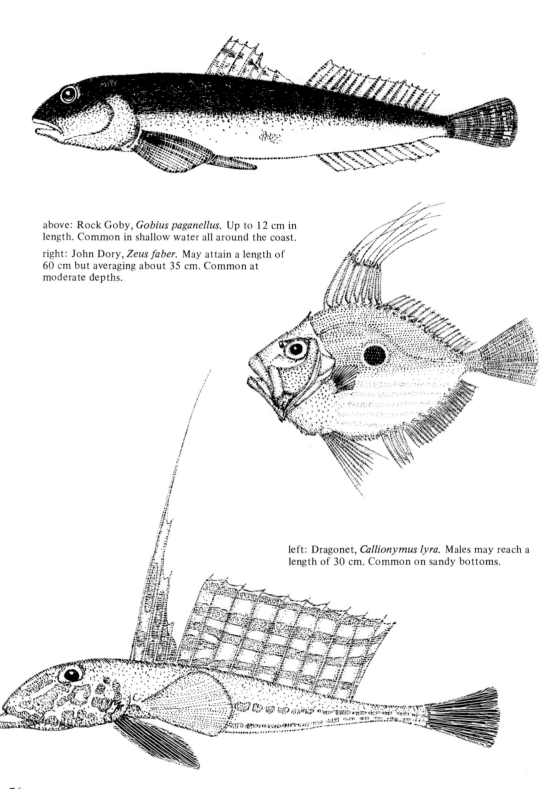

above: Rock Goby, *Gobius paganellus.* Up to 12 cm in length. Common in shallow water all around the coast.

right: John Dory, *Zeus faber.* May attain a length of 60 cm but averaging about 35 cm. Common at moderate depths.

left: Dragonet, *Callionymus lyra.* Males may reach a length of 30 cm. Common on sandy bottoms.

above: Tompot Blenny, *Blennius gattorugine.* Length about 16 cm. Not uncommon beyond low tide mark. Occurs mostly off South-western coasts.

below: Streaked Gurnard, *Trigloporus lastoviza.* Recorded only occasionally off the South coast.

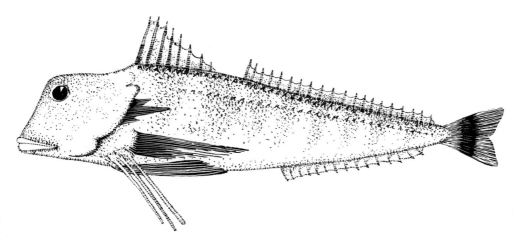

below: Montagu's Sea Snail, *Liparis montagui.* Generally about 10 cm in length. Common in seaweed zone just below low tide mark.

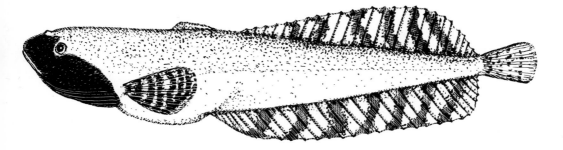

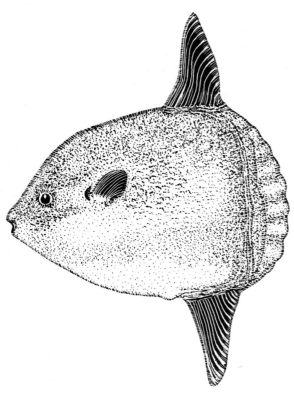

above: Sunfish, *Mola mola.* Enormous specimens of
3 m in length are known but are usually only
about 1 m. Occasionally reaches Irish waters.

right: Angler, *Lophius piscatorius.* Generally about
50 cm in length. Widely distributed around the
coast and found on various types of bottom down to
1,000 m.

below: Connemara Sucker, *Lepadogaster candollei.*
About 6 cm in length. Recorded from a few localities
on the West coast.

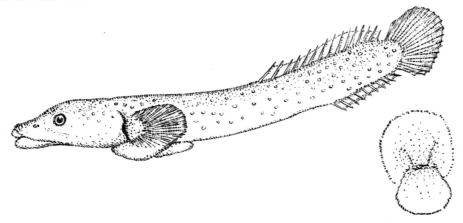

Suctorial Disc

78

common. In both species the dorsal and the ventral fins are very long, but whilst in the case of the Common Sea snail, they both connect with the tail fin, in Montagu's they do not quite connect.

The abdomen appears swollen, giving this fish a rather tadpole-like appearance. The skin is slimy and scaleless and the ventral fins are modified to form a sucker. As Sea snails occur in the seaweed zone, just off low-tide mark (where wave activity is general), the value of being able to adhere to a nearby rock or stone can be appreciated. The colour of the species varies with the colour of the seaweeds amongst which they are to be found. They may be yellowish, brown or reddish. Montagu's Sea snail averages about 10cm in length, whilst the Common species is a little larger. They feed on small crustaceans.

Sunfish *Mola mola*

This most unlikely-looking fish, of tropical and temperate regions, from time to time, is found in Irish waters. In appearance it is almost plate-like with oar-like dorsal and anal fins which are only endowed with side to side movement. Capable of drifting on its side or swimming upright near the surface, the dorsal fin is often clear of the water and waving to and fro. This, in fact, distinguishes it from sharks, as seen from the deck of a ship. There is no tail fin, this being replaced by a wavy outline of stiff skin. The pectoral fins are small and are situated in the middle line, not far from the eyes. Ventral and pelvic fins are also absent.

The scaleless skin is extremely thick and leathery with a tough fibrous layer immediately beneath — presumably as a protection from large predators. It attains an enormous size — specimens of 3m in length and a weight of 1,400kg being known. Generally however, sunfish occasionally reaching Irish waters are less than a third of this size. The mouth is small and it feeds on planktonic animals. Specimens at the National Museum were collected off Achill Island, Co. Mayo, in 1873, and Lough Swilly, Co. Donegal, in 1888. It is closely related to the Truncated sunfish, *Ranzania truncata,* which is similar in some respects to the above species,

but has a more elongated body. The Basking shark is also called a sunfish.

Angler *Lophius piscatorius*

Even the most ardent fish-lover could scarcely describe the angler as beautiful. In shape it is squat, the head and forepart being circular and behind the fleshy, square pectoral fins the body tapers narrowly. A large part of the head is taken up by the mouth and its backwardly-directed teeth are needle-sharp.

The back is mottled or reticulated with various tones of brown and green, whilst underneath it is white. The most remarkable feature, however, of this bottom-dwelling fish concerns the dorsal fin, the first three spines of which are separate. The first spine can be reclined forwards, and at the apex is a tassle which can be moved to simulate a live organism! This attracts fish as being possible prey but, when within reach of the angler's jaws, it is snapped up with a lightning-quick movement.

The Angler has a wide distribution around the coast from muddy or sandy bays down to depths of 1,000m, and the violet-coloured egg-band of about 10m in length and containing about one million eggs is yet another remarkable feature. The adult fish measures from 40 to 60cm on average but sometimes a very large size is attained. The record weight of an angler taken from Irish waters by fair angling is of 32.43kg caught in Cork Harbour in July 1964.

Connemara Sucker
Lepadogaster candollei

The family of suckers, GOBIESOCIDAE, is almost exclusively tropical in distribution, but four species in one genus inhabit Irish shores. Two of these are found only on south and west coasts of Ireland, their main distribution being Mediterranean. The Cornish sucker, *Lepadogaster lepadogaster,* occurs on the south and south-west coasts of both Ireland and England, whilst the Connemara sucker, *Lepadogaster candollei,* is found only at a few localities on the west coast. The other two species are most widely distributed.

All species are small, the Connemara sucker

illustrated was only 6cm in length. They all possess a suctorial disc on the belly, immediately beneath the pelvic fins — indeed, the disc is made up of a prolongation of the lower part of these fins, together with a cartilaginous extension of the coracoid bone of the pelvic girdle. The elongated eggs are laid in empty mollusc shells and hatch in about 4 weeks. In the first winter, the young fish migrate into deeper water, but thereafter return to the intertidal zone where they are to be found under stones in small, tidal pools. The possession of the sucker enables the fish to resist being washed about and buffeted during tidal movement.

The Connemara sucker is easily separated from the other three species in the genus by having the dorsal fin distinctly differentiated from the caudal fin, and by the possession of more than ten rays in the dorsal fin.

6 Molluscs

Bivalve Molluscs, Snails, Slugs, Sea Slugs, Squids

MOLLUSCS *Mollusca*

Molluscs are the second largest group of animals and comprise some 100,000 species. Most of these are marine, some freshwater and comparatively few are terrestrial. Of the 650 species to be found in Ireland and Britain about 450 are marine. There are no significant differences between the Irish and British marine molluscs. Of the remaining 200 species about a quarter have not been recorded from Ireland, and in general they are extremely rare in Britain. A few species occur in Ireland but not in Britain, such as the Kerry slug *Geomalacus maculosus*.

The molluscs consist of three major classes — the slugs and snails, etc., the *Gastropoda*; the scallops, cockles and mussels, the *Lamellibranchia*; and the squids and octopuses, the *Cephalopoda*. There are also several smaller classes — small, worm-like, primitive molluscs, the *Solenogastres*; the chitons or coat-of-mail shells, the *Loricata* and the Tusk shells, the *Scaphopoda*. Clearly the *Mollusca* is a highly diverse group but all are characterized by their soft, mucus-coated bodies; their possession of a hard, calcareous shell during some stage of their life and the frequent possession of a radula, a long, looped tongue bearing numerous rows of chitinous teeth.

Geologically this is an ancient group, first appearing in pre-Cambrian times some 600 million years ago, and it is of importance in that the hard shells have been preserved as fossils, and the knowledge of their evolutionary development is well-advanced.

Proneomenia Aglaopheniae

This primitive mollusc is worm-like in appearance and is generally found coiled around the stalk of the hydroid *Theocarpus myriophyllum* (formerly called *Aglaophenia*). It occurs at a depth of 50 to 100m.

Smooth Chiton *Callochiton achatinus*

The Chitons or 'coat-of-mail' shells constitute the Class *Loricata*. They are easily identified by their possession of a shell divided into 8 plates arranged in a row surrounded by a fleshy girdle — *Loricata* means 'corset-bearing'. They are thus able to conform in shape to the curved surfaces of a stone or a rock to which they adhere. Many tropical species reach a length of 12cm or more but the Irish species seldom attains 2cm, although *Tonicella marmorea,* which may be found as far south as Dublin, has been recorded up to 4cm in length. The Smooth chiton has reddish plates which are marked with white, and the girdle is bordered by very small spines. It is about 2.5cm in length, and may often be found at low water under stones, and down to a depth of 150m.

Common limpet *Patella vulgata*

The *Gastropoda* (meaning stomach-foot) is an important division of the *Mollusca*. It includes both aquatic (marine and freshwater) and terrestrial species, and whilst most carry obvious shells, some do not. Well-known gastropods are limpets, winkles, whelks, snails, Land slugs, as well as Sea slugs with many others besides. The essential characters are those of organization. The body is distinctly separated into a head and foot (or sole) region, with organs of sense situated in the head. The eyes, set on long, tentacle-like stalks are an example of this. Feeding and locomotion concern the head-foot region, whilst the visceral organs are situated on the back — often as a coil at the distant end of a spiral shell.

The body is normally divided into a foot bearing distinct head and a visceral hump covered by the shell. The major sense organs are situated on the head. The eyes are situated either at the tip or the base of the head tentacles which bear the organs of taste and smell. The head and foot regions are also concerned with locomotion, whilst the internal organs, such as the gut, liver, heart and kidney, are tucked up in the visceral mass often in the tightly coiled shell.

The limpet is able to close access to the inside of its shell by pulling down tightly, by suction, onto a smooth rock surface to which it fits closely. It is most difficult to dislodge. The limpet browses on algae growing on the adjacent rock surfaces at low tide at night and returns to its exact resting place before the tide flows in. There are eight species of true limpet to be found in Irish coastal waters. The Common limpet is generally about 5cm in diameter but may attain 7cm. It is an abundant inhabitant of all rocky shores between tide-marks.

Cowrie *Trivia monacha*

The cowrie shell is one of the best known of shell forms, and many local names are used for it. Two species classified in the family ERATOIDAE occur in various localities around the Irish coast. The Spotted cowrie, *Trivia monacha,* has three

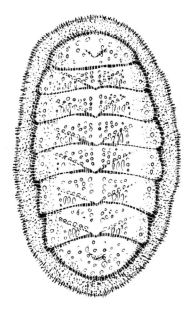

above: Smooth Chiton, *Callochiton achatinus.* Maximum length 25 mm. Probably common from low water mark to a depth of about 150 m.

right: A primitive mollusc, *Proneomenia aglaopheniae.* About 20 mm in length. Probably widespread at depths of 50-100 m wherever the hydroid *Theocarpus myriophyllum* occurs.

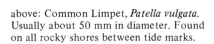

above: Common Limpet, *Patella vulgata.* Usually about 50 mm in diameter. Found on all rocky shores between tide marks.

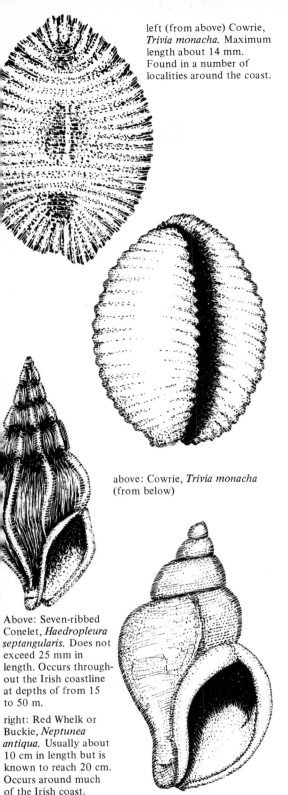

left (from above) Cowrie, *Trivia monacha.* Maximum length about 14 mm. Found in a number of localities around the coast.

above: Cowrie, *Trivia monacha* (from below)

Above: Seven-ribbed Conelet, *Haedropleura septangularis.* Does not exceed 25 mm in length. Occurs throughout the Irish coastline at depths of from 15 to 50 m.

right: Red Whelk or Buckie, *Neptunea antiqua.* Usually about 10 cm in length but is known to reach 20 cm. Occurs around much of the Irish coast.

brownish-purple spots on the back of the shell, which reaches a length of 14cm. The body is orange in colour and the mantle partly envelopes the strongly-ribbed shell. The species, *Trivia arctica,* does not possess spots and is usually smaller than *T.monacha.* It also has a wider distribution and is generally more abundant. The third species, the scarce *Erato voluta,* also occurs in Ireland. Its smooth, milk-white, porcelain-like shell has about fifteen small teeth within the outer lip.

Red Whelk or Buckie
Neptunea antiqua

The buckie is about 10cm long and the shell is strong and thick, being yellowish or pinkish in colour with numerous, fine spiral ridges. It is extremely variable both in colour and form — sometimes it reaches a length of 20cm. This species is edible. The buckie is a northern species but is found around much of the Irish coast — generally inhabiting the coralline zone.

Whelks are carnivorous gastropods characterized by the structure of the radula. Each row of teeth consists of a central tooth with five to seven stout spines and lateral teeth with two or three cusps. In common with other marine carnivorous snails there is a tube-like forward extension of the shell. This extension guides the siphon which is held in front of the animal and is used to detect the smell of potential prey.

Seven-ribbed Conelet
Haedropleura septangularis

The conelets are related to the cone shells of tropical waters, some species of which are venomous, and a number of fatalities are known amongst collectors handling them. All our conelets, however, are small, never exceeding 2.5cm in length. They are carnivorous in habit and are characterized by the cone-shaped shell, narrow mouth and a great reduction in the number of teeth on the radula. (Teeth can also occur in the mouth of the shell.) The Seven-ribbed conelet is reddish-brown in colour with seven pale, somewhat sinuous ribs which cross the body-whorl. The pear-shaped operculum is amber coloured. This species occurs around the whole Irish coastline at a depth of from 15 to 50m.

Canoe Shell *Tricla*
(formerly known as *Scaphander*) *lignaria*
The bubble shells, the family RETUSIDAE, have a very light shell with the aperture extending the whole length of the shell — hence the name of this species. It is reddish-yellow and marked with fine lines and grooves which give it a woodgrain appearance, and the body of the animal is yellowish with a square-shaped, eyeless head and heel-like lateral lobes which enable it to burrow into sand. The tentacles form a square lobe behind the head. The animal reaches a length of about 5cm. It possesses a very strong gastric mill which enables it to crush shells of other molluscs on which it feeds.

Although it feeds on a wide variety of animal organisms, it appears to hunt specifically the tusk shell, *Dentalium entalis.* An association of the Hermit crab, *Eupagurus prideauxi,* inhabiting the empty shells of the canoe shell itself, and the Sea anemone, *Adamsia palliata,* which envelopes the back of the shell, sometimes occurs. The canoe shell is found at depths of from 100 to 200m, and is fairly common.

Sea Hare *Aplysia punctata*
The Sea hares are large, soft bodied animals with the horny shell completely internal. Presumably there is only one Irish species, as given in the title, but there are always personal doubts as to the identity of the large specimens which the author has observed in Dunmanus Bay, Co. Cork. It has also to be admitted that the author has, so far, refrained from cutting up one in order to examine the shell!

The Sea hare is a remarkable animal. The shell is a flat, transparent, flexible plate buried beneath the integument. The head bears two pairs of tentacles and is connected by a rather long neck with the main body. Two large lobes extend from the foot and generally curve over the back. A siphon is formed from a fold continued backwards from the mantle covering the shell. Overall it reaches a length of 15cm. When it considers itself menaced it secretes a purple fluid from an organ at the edge of the mantle. It is said to cause an indelible stain and it was named *Aplysia* — meaning unwashable!

Perhaps the most remarkable phenomenon exhibited by *Aplysia punctata* is the correlation of its colour with the seaweeds on which it feeds at different stages in its life. When young, the Sea hare is bright crimson, spotted with white — almost corresponding exactly with the red seaweed, *Delesseria,* on which it then feeds. Later it feeds on dulse *(Iridaea)* when it then matches the deep red-brown colour. As time goes on it may be found amongst the darker seaweeds — purple-brown, olive-brown, olive green — all of which, at some stage, it always seems to match! Although most descriptions refer to it as 'spotted' (as does its specific name), frequently spots are absent — as in the Dunmanus Bay specimens.

Plumed Aeolis *Aeolidia papillosa*
Sea-slugs are soft-bodied molluscs entirely lacking shells and known as the true Sea-slugs. Many of the species are clothed, to a greater or lesser degree, with long, fleshy, finger-like but pointed protrusions known as 'cerata'. The latter contain components of the digestive gland, but, in addition, a number of species possess tentacles associated with sense organs. Nudibranchs are carnivorous and feed upon sponges, sea anemones and related organisms.

The Plumed aeolis is fairly large, often reaching up to 12.0cm length. The over-lapping cerata are greyish or brownish, purple or white. It is fairly common, and one species is said to mimic sea anemones, even to the extent of possessing stinging organs on the tips of the cerata. (These have been transported from the prey undischarged). If the cerata are lost, due to a predator browsing upon them, they are regrown. It is believed, however, that they are generally unpalatable to fishes.

Rough Sea Lemon *Doris tuberculata*
This species is usually of some considerable size — about 75mm in length but sometimes as long as 125mm. The animal is completely enveloped by the mantle which is covered with small

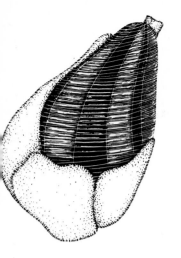

above: Sea Hare, *Aplysia punctata*. Reaches a length of 15 cm. Occurs not uncommonly in the seaweed zone.

left: Canoe Shell, *Tricla lignaria*. Attains a length of about 50 mm. Fairly common at depths of from 100-200 m.

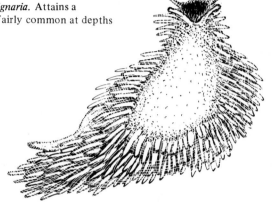

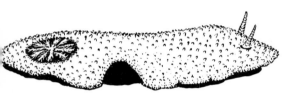

above: Rough Sea Lemon, *Doris tuberculata*. About 75 mm in length but sometimes up to 125 m. Found wherever the Bread Crumb Sponge, *Halichondria panicea* is abundant.

above: Plumed Aeolis, *Aeolidia papillosa*. Often attains a length of 75 mm. Fairly common.

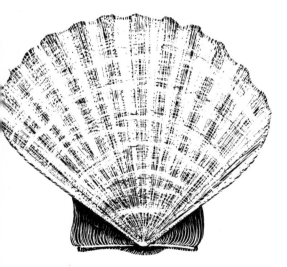

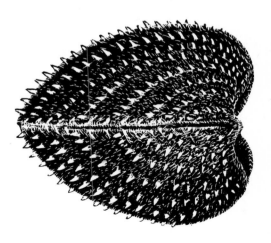

above: Great Scallop, *Pecten maximus*. May reach 120 mm in diameter. Common.

above: Prickly Cockle, *Cardium echinatum*. Up to about 50 mm across. Fairly common on sandy bottoms at depths from 10-200 m.

tubercles. A pair of tentacles emerge through slits in the mantle, and at the other end is the anal cavity in which there is a rosette of branched gills. In colour it mimics the Breadcrumb sponge, *Halichondria panicea,* on which it feeds, the colour being pale yellow to orange, clouded with grey, pink and green. If the Rough sea lemon is removed from the sponge it is immediately conspicuous, but when it browses upon it, it is almost invisible.

Great Scallop *Pecten maximus*
The oysters, scallops and file shells, as well as the Fan mussel, are classified together (order *Pseudolamellibranchia*). The Great scallop is the largest of the nine scallop species, and may be up to 12cm in diameter. The upper valve is practically flat, but the lower one is deeply convex. The shells are deeply ribbed. This not only gives added strength, but the two valves fit so accurately together that sideways movements of the valves is effectively prevented. There are wide 'ears' on each side of the beak. The ligament is largely restricted to the region not containing the 'ears'. The function of the ears is probably to direct the water currents during swimming.

Usually the shell is open slightly when the mantle is seen to be double with the outer one, bearing a number of brilliantly coloured opalescent eyes at the base of the fairly long tentacles. The inner mantle is fringed. A finger-like foot protrudes from between the ears.

All species of scallop are able to swim in a zig-zag course by clapping the two shells together — that is except one species, the Hunchback, *Chlamys sulcata,* which lives permanently attached, by means of threads, to rock cavities. There are two rare species — *Chlamys nivea* (found near Glengarriff, Bantry Bay, County Cork) and *Chlamys furtiva,* found in Bertraghboy Bay, County Galway, and at Larne, County Antrim.

Prickly Cockle *Cardium echinatum*
Eight Irish species of cockle comprise the family CARDIIDAE. The latter is derived from the shape of a heart which the closed shell resembles

left: Pod Razor, *Ensis siliqua.* Maximum length of about 20 cm, width about 3 cm. Common and generally distributed in sandy localities.

below: Wrinkled Rock – Borer, *Hiatella arctica.* Up to 35 mm in length. Common around the coast where 'softer' rocks such as limestone occur.

left: Paper Piddock, *Pholadidea loscombiana.* About 37 mm in length. Uncommon, with chief locality being Co. Antrim coast.

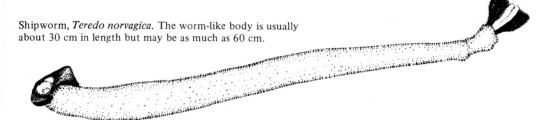
Shipworm, *Teredo norvagica*. The worm-like body is usually about 30 cm in length but may be as much as 60 cm.

when viewed from the side. The two valves are strongly convex and exactly alike. Radiating from the prominent beaks are strong ribs and their adjacent furrows. These continue to the shell-margins which interlock accurately with each other. In colour it is yellowish-white with a little reddish tinge, and it is up to 50mm or so across. The ribs bear short, backwardly projecting, triangular spines. There is a finger-like foot about twice the length of the shell. It is a fairly common species, and is found on sandy bottoms from 10 to 200m.

Pod Razor *Ensis siliqua*
The razorfish and spoutfish constitute the Family SOLENIDAE. These are characterized by the long, narrow, equal-length shells which gape at each end and have a strong foot, which enables it to speedily burrow into sand. It reaches a length of about 20cm and a width of about 3cm. The exterior, fine ridging shows that the main addition in growth is on one side only. The united siphons are very short and fringed at their openings, and pigment spots or 'eyes' are present which are sensitive enough to give the signal for alarm and initiate burrowing if a shadow is cast over the animal when exposed — that is, when the tide is in. When the tide is out, it burrows diagonally beneath the sand for about 30cm or so. It is of interest that it can swim by means of water ejection. A rare, related species, *Ensis minor,* only half the size of the above species, has been recorded from Newcastle, County Down.

Wrinkled Rock-Borer *Hiatella arctica*
There has been some confusion in the past over the naming of this species. Some older books refer to it as *Saxicava rugosa.* Because it spends

its life boring into rocks — limestone, chalk, or new red sandstone — the casual shell collector seldom finds it. However, if small holes are noticed in such rocks and the tunnels exposed, then the somewhat distorted shells of these molluscs may be found. Additionally, however, it may be found nestling into *Laminaria* (oarweed) holdfasts, thereby extending its range into shores containing basalt and granite. The shells are thick and dull and up to 35mm in length — a rather dirty-white colour. The siphons can be extended considerably as they obtain no nourishment from the rocks which provide only shelter, so that small organisms are filtered out of the sea water by means of the siphon. The Wrinkled rock-borer is common and found around the coast wherever the suitable rock-types occur, from low water mark down to a depth of about 60m.

Paper Piddock *Pholadidea loscombiana*
The several species of piddock constitute the family PHOLADIDAE and are well-known boring molluscs. The Paper piddock (it was once known as *papyracea*) is easily identified when an adult, by the horny cup encircling the base of the siphons. The shells are exceptionally thin in the forepart. It makes its burrows in sandstone, submerged peat and wood as well as hard clay. Known from several localities, chief of which is the Antrim coast.

Shipworm *Teredo norvagica*
Almost everyone must know of the shipworm, which ate away the wooden hulls of sailing ships and often caused devastation in tropical seas. Few would know it to be a mollusc, even when taken from the partially destroyed timber. The body is wormlike and generally grows to about

30cm in length, but can be up to 60cm in length. The shells touch only at the beaks, and at a point on the margin opposite and they are three-lobed. At the other end of the animal are the siphons which are exceptionally long when extended. There is also a pair of shell-like 'pallets' – in the shape of tennis racquets!

The animal bores into the wood and in so doing a hard, shiny coat of calcium carbonate is secreted onto the walls of the tunnel. The latter becomes larger as the boring progresses, whereas the entrance hole remains small, and only sufficiently large enough for the siphons to emerge in order to collect minute organisms for food, oxygen for respiration and for the excretion of faecal matter. The major food source, however, is derived from the timber. It is locally common in wooden piers and generally to be found in harbour works but a related, smaller species, *Teredo navalis,* is probably more numerous.

Common Squid *Loligo forbesii*

The most highly developed group of *Mollusca* is the *Cephalopoda* (meaning head-foot), which comprises the cuttlefishes, squids and octupuses, characterized by their possession of head-tentacles furnished with suckers, a strong, hard beak like that of a parrot, very efficient mammal-like eyes and large brains. The shell is usually internal, but in the *Nautilus* it is external. All species are carnivorous, and are able to effect their escape from larger predators by fast, jet-propulsion – forcing sea water backwards from a powerful, muscular, bag-like organ, and emitting a 'smoke-screen' of black, ink-like liquid in which they can hide. About twenty species occur in Irish waters. Very rarely, giant species of Cephalopod (about 10m across the outstretched tentacles) are washed ashore and these are probably from oceanic, deep waters. When this occurs, the nearest museum should be notified at once.

The Common Squid is widely distributed in Irish water and varies in size from 20 to 75cm but averages 30cm, being torpedo-like in shape with triangular side-fins. The internal shell consists only of a horny pen. Cephalopods are divided into two main groups – the *Decapoda* which have five pairs of tentacles (one pair longer

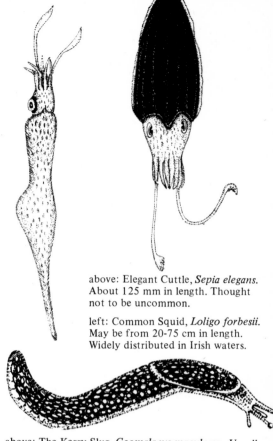

above: Elegant Cuttle, *Sepia elegans.* About 125 mm in length. Thought not to be uncommon.

left: Common Squid, *Loligo forbesii.* May be from 20-75 cm in length. Widely distributed in Irish waters.

above: The Kerry Slug, *Geomalacus maculosus.* Usually about 5.5 cm in length but may reach 9 cm. Found only in a few areas in Counties Kerry and Cork where it may be locally abundant.

above: Banded Snail, *Helix nemoralis.* About 25 mm in diameter. Common throughout Ireland. (left) Dark form. (right) Light form.

left: Little Cuttle, *Heterosepiola atlantica.* Usually from 25-50 mm in length and generally common on sandy coasts.

than the others), and in this group are the squids and the cuttlefishes. The only members of the other groups — the *Octopoda*, which have four pairs of tentacles all of the same length — are the octopuses.

Elegant Cuttle *Sepia elegans*
The length of this cuttle is about 12.5cm which is much smaller than the Common cuttle, *Sepia officinalis*, which produces the well-known cuttlefish 'bone'. The latter species may reach a length of 45cm. The Elegant cuttle is proportionately narrower than the Common species and is often pink in colour. The specimen illustrated was a rich, carmine colour.

Little Cuttle *Heterosepiola atlantica*
This species is only about 2.5 to 5.0cm in length, and has a rounded body, large eyes and lateral, lobe-like fins. It burrows into sand by directing its water-jet at a spot until the sandgrains have been cleared enough for it to bury itself. The Little cuttle is usually common on sandy coasts and is frequently taken in shrimp nets.

Banded Snail *Helix nemoralis*
The family HELICIDAE contains a large number of terrestrial species, some of which are exceptionally common. The Common garden snail, *Helix aspersa*, is one such, and is to be found practically everywhere in gardens except in industrial areas. The Banded snail is sometimes called the 'Brown-lipped' snail to distinguish it from the White-lipped snail, *H. hortensis*, which it greatly resembles. This species is more or less globular in shape and reaches a diameter of about 2.5cm. It is exceptionally variable because, although the ground colour can generally be described as yellow with five dark bands following the whorls, the dark bands often almost obliterate the ground colour and coalesce into each other. Most books on European molluscs give much illustration space to the almost endless variety of forms. The Banded snail is found throughout Ireland.

The Kerry Slug *Geomalacus maculosus*
There is some difficulty in considering a slug as a treasure of nature! Slugs are not attractive creatures, indeed, most people are repelled by them. Perhaps this is due to their slow and slimy progress or, perhaps, to their destructive powers as they eat their way into our vegetables and flowers in the garden. It must be stated, however, that of about thirty species of slug only about six are significant pests. The remainder feed largely on detritus, lichens and fungi. The Kerry slug, *Geomalacus maculosus*, is different. This is not because it is unlike other slugs, but because to those of us interested in the distribution of animals, it poses exciting questions!

Let us look at this distribution. The Kerry slug occurs only in two areas — the extreme southwest of Ireland (in Kerry and Cork), and in a triangular area in the northwestern corner of the Iberian peninsula in Portugal or Spain. It also occurs in Verres in France. Thus our slug is one of the species making up the Lusitanean fauna.

The Kerry slug can be differentiated from the various species of the genus *Arion* by the genital orifice being behind and below the right upper tentacle. The 'shell' is a compact, oval, calcareous mass, and the spotted colour pattern is also diagnostic. The length, when fully extended, may be as much as 9cm, but is usually only about 5.5cm. The mantle takes up about one third of the body length and is elliptical in shape, and the respiratory orifice is located a little in front of the centre right margin. The colour is variable, from grey to almost black with irregular whitish or yellowish spots. Many of the colour varieties have been named. There is an internal shell which is oval and measures about 4x3mm.

An important behavioural characteristic, which serves to distinguish the Kerry slug from its near relatives in the genus *Ater*, is its habit of rolling up into a ball if disturbed — somewhat like the woodlouse *Armadillidium*. The Kerry slug is to be found feeding on algae, mosses and liverworts which coat the boulders on the mountain slopes of Old Red Sandstone.

Arthropoda, an Introduction

The *Arthropoda* is one of the most important groups of animals. It is certainly the largest in numbers of species and exercises an effective control over the numbers, not only of all other animals, but also of many plant species. Arthropods exhibit the following characteristics which should be interpreted by the student in a general manner. The body is elongate with mouth and anus at opposite ends, and it is divided into a number of segments in which there is some serial repetition of organs and limbs. They are bilaterally symmetrical, and the body is encased in an external skeleton of the substance chitin, which may be impregnated to a greater or lesser extent with calcium. The body segments and the limbs are moveable so that the hard, sclerotized segments alternate with flexible membranes. The most important sub-divisions of the *Arthropoda* are as follows:

Arachnida	This includes spiders, mites and ticks.
Insecta	This includes flies, beetles, butterflies, moths, aphids and bees.
Myriapoda	This includes centipedes and millipedes.
Crustacea	This includes crayfish, crabs, shrimps, woodlice, barnacles and water fleas.

These are dealt with in Chapters 7 to 12.

7 Spiders

Spiders, together with scorpions, harvestmen, mites and ticks, as well as a few lesser-known groups, constitute the class *Arachnida*. Members of this class differ from insects in having the body divided into two main parts instead of three, as in insects. Arachne was the name of a maiden in mythology who was transformed into a spider by the goddess Athena. All spiders are predaceous — mainly preying on insects — and best known from the habit of many species of spinning webs in order to snare their prey.

In general organization the spiders do not exhibit such clear segmentation as do insects. The head and the thoracic regions are fused together and are known as the cephalothorax or prosoma, whilst the abdomen is often referred to as the opisthosoma. The cephalothorax carries six pairs of limb-like, jointed appendages. The first pair lie immediately in front of the mouth and are known as chelicerae (having two or three segments), and function as jaws. The second pair are tactile organs, known as the palpi or pedipalpi, and these are followed by four pairs of 'walking' legs which are jointed, covered with various kinds of spines and hair — many of which possess sensory functions. The tips of the legs each have two or three tough, sharp claws. The abdomen is markedly divided from the cephalothorax by a waist — the pedicle. Although often not seen clearly from the outside, it is composed of ten 'somites'. None have appendages, but the silk or gossamer for web-making is produced by the spinnerets, which are highly modified appendages borne by the fourth and fifth somites.

Looking at the front of the spider from the top, the eyes, numbering eight or sometimes six are seen as simple, black dots. Their various arrangements are of importance in classification. At the front, but underneath are the chelicerae, which are modified into poison fangs. They consist of two parts: the broad basal joint which is hollow and encloses the poison duct, and the apical joint which has a hard, sharp claw through which the poison duct opens at the tip. The palpi are six-jointed, and the sex of the adult may be determined by examining the tarsus — the last joint. In the female this bears a claw but in the male it consists of a sex organ of complicated construction.

The process of mating in spiders is unique in the whole of the animal kingdom. When the male becomes sexually mature he spins a small patch of silk on which he deposits a drop of seminal fluid containing sperms. This is then sucked up by the highly modified 'pedipalps' which then perform the function of a penis, and the parcel of sperm is inserted into the vagina of the female. This is, however, not before an elaborate courtship has taken place which has the effect of immobilizing the voracious jaws of the female.

There are seven joints to the legs, and these are named commencing from the body: coxa, trochanter, femur, patella, tibia, metatarsus and tarsus, the claws at the tips of the tarsus are comb-like. The spider's mouth is very small but a number of auxilliary parts surround it. The palpi are borne by plates known as maxillae, between which lies the lip (labium). The maxillae squeeze the juice from the prey, which is then sucked into the body via the pharynx. Still viewing the underside of the cephalothorax, a large conspicuous plate, oval or heart-shaped, is observed. This is the sternum, and on the abdomen behind it (and near to the waist) is a pair of pale areas which are the lungs. Between these lie the opening of the oviduct (epigyne).

In some families, there is an additional spinning organ known as the cribellum. It has the appearance of an oval sieve, and silk is combed out of it with a row of spines known as the calamistrum on the fourth pair of legs. Close to the epigyne is a pair of apertures leading to the spermathecae — the reservoirs for storing sperm until egg-laying starts. The complexity of these structures varies greatly and is again important in identification. Further details of structure and behaviour of spiders will be given in the descriptions of the species described in this chapter.

The Crab Spider *Xysticus cristatus*

Members of the family THOMISIDAE are known as Crab spiders on account of their habit of running backwards and sideways, in addition to forwards, such as does a crab. They do not spin webs, nor do they chase their prey, but secure prey by waiting in a favoured situation where insects may alight, and this is frequently a flower. These spiders often show colouration similar to that flower in which they hide, so that the unsuspecting insect may be caught by a sudden pounce. Other species are coloured in various browny shades, and hide amongst dead leaves, awaiting their prey. Some confusion may arise over the naming of the species illustrated. It has also been given the name of *Thomisus cristatus* and *Xysticus viaticus*. However, it is the most abundant of the Crab spiders.

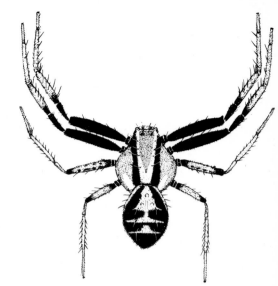

above: Crab Spider, *Xysticus cristatus*. Body 8-10 mm in length. Most abundant of the Crab spiders. Male.

below: Crab Spider, *Xysticus cristatus*. Female.

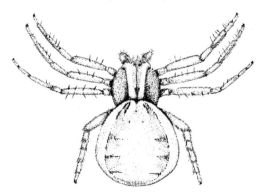

The body is about 8-10mm in length, and the general colour is a mixture of yellowish-browns, but the patterning may be nearly black — the variable background colour may be green — as in the illustration. The two front pairs of legs are carried forwards, closely parallel with each other as in most species. The cephalothorax carries a pair of backwards-converging lines, between which a triangular mark has an apex pointing backwards. The pattern on the abdomen consists of a series of overlapping triangular marks, more or less conspicuous. In the more darkly-coloured male, illustrated with a small abdomen, they are easily seen, but in the illustration of the female with an expanded abdomen, they are not so evident.

Pholcus phalangioides

This species is the only member of the family, the PHOLCIDAE, found in Ireland, and is perhaps the most easily identified of all Irish spiders. This is on account of its extremely long, thin legs, black at the joints; a circular cephalothorax 2mm in diameter; a pale yellowish-brown to grey cylindrical abdomen 6mm in length, and a black band along the median line. When resting, the legs stretch out to about 5cm across and the same distance from front to back. No other Irish spider *remotely* resembles it. The novice, however, must be on guard against confusing it with a harvestman (see later), but the division of the body — very definitely divided into a round cephalothorax and a long, bag-like abdomen — most surely identifies it. A microscopic feature of interest is the tooth-like projection on the inner-side of each of the chelicerae. The fang which is jointed can be opposed to it, thus making it a pincer.

This spider is not known out-of-doors. Its web is constructed haphazardly, and forms a tangle of random threads which are always near the ceiling from which the spider hangs, head downwards, in the middle. A curious habit for the *Pholcus* also is that of vibrating at great speed when a fly touches a web-thread. This serves to entangle the prey still further, and the speed of the vibrations is such that the spider is almost invisible. The author has often wondered if this may not be, in addition, a defence mechanism. The twenty or so eggs of this species — in the slightest of silk cocoons — are carried about by the female in her *jaws,* and on hatching the young hold onto her face for a time, and then run off to fend for themselves.

Gorse-Web Spider *Agelena labyrinthica*

The family AGELENIDAE contains perhaps the most well-known species of all the common house-spiders which leave their webs around windows and other corners in our dwellings. Additionally, a number of species are common out-of-doors.

A characteristic of the family, easily seen with the naked eye, is the length of the first pair of spinnerets which trail behind the abdomen like a pair of tails. The webs of the Gorse-web spider are common sights in the countryside wherever gorse is to be found. These are large 'sheets' of web, converging into a tunnel, and the reason for the choice of gorse as a site for the web is because this spider prefers a dry situation, which, apparently gorse provides.

The length of the specimen illustrated was 14mm, but usually this species is a few mm less. The cephalothorax is bright brown in colour, with a dark-brown or almost black longitudinal band on each side. The large, oval abdomen is black with a series of light-coloured, transverse bars — each with a forward prolongation in the mid-line. The legs are dark-brown in colour, stretching to about 23mm from front to back when at rest. During the mating season, the sexes live amicably together. A large, silken egg-cocoon is produced in July.

Oval Bush Spider *Enoplognatha ovata*

The family in which this species is classified, the THERIDIIDAE, consists of mostly small spiders, but generally also brightly-coloured. Their lives are spent just a few feet above the ground on shrubs or in hedges. The web is of the simplest construction made up of only a number of criss-crossing threads — being almost invisible. This spider spins a silk-lined shelter for itself amongst a few leaves fastened together with more silk and having access to the web. In *Enoploghatha ovata*, the length of the body is about 7mm overall, when resting. The cephalothorax is almost black and the eyes are elevated on relatively large tubercles. Its legs are brown. The oval abdomen has a cream-coloured background with a black border and three discrete, triangular marks with the apices pointing forwards, but the first of these has a round mark surmounting it.

The Harvestman *Mitopus morio*

The harvestman constitutes the separate order *Opiliones* of the *Arachnida.* They are easily separated from spiders and other arachnids by the following characters: firstly and *most* important, there is no clear distinction or 'waist'

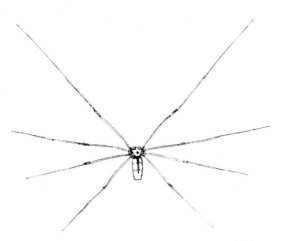

above: *Pholcus phalangioides.* When at rest with legs stretched out it measures 5 cm in each direction. Not known out of doors. Specimen illustrated is from West Cork.

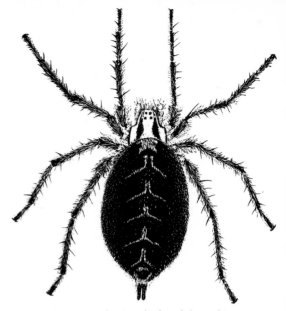

above: Gorse-web Spider, *Agelena labyrinthica.* Length 14 mm. Common wherever Gorse is found.

below: The Harvestman, *Mitopus morio.* Length of female 8 mm, male 5 mm. Widely distributed and common throughout Ireland.

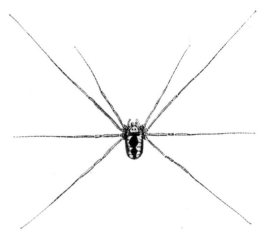

between the cephalothorax and the abdomen. However, like spiders, they do have six pairs of legs (appendages), the last four being four walking legs attached to the cephalothorax, and these are always extremely long and thin in relation to its 'dumpy' body. *Mitopus morio* is the species taken to illustrate and describe an example of a harvestman, found in Ireland. It is relatively easy to identify, but the sexes are fundamentally different in size and colour.

The female is about 8mm in length and the male 5mm. The female is a creamy colour marked with dark-brown or black, and the male is very much darker, often almost black, the most significant difference being the 'hour-glass' pattern of the female. On the upper surface, in both sexes, the cephalothorax has an ocular tubercle which carries a pair of simple eyes, each facing sideways. Close to the basal segment of the second pair of legs there are stink glands, often conspicuous, and not to be taken for 'eyes'. These glands secrete a nauseous liquid with which the harvestman may annoint itself — thus, perhaps, dissuading a predator from devouring it! In front of this ocular tubercle there is a pattern of spines — the trident. Some species carry quite large tridents, but in *M.morio* it is small, and the chelicerae are made up of three strong segments, forming pincers because of a prolongation arising from the second segment reaching to the tip of the third. The chelicerae do *not* contain poison fangs.

M.Morio is widely distributed and is common throughout Ireland — even as far up as the summits of mountains. It occurs from North America eastward to China. It is the only species of harvestman to be found in Greenland. Another large harvestman found in the same situations as *M.morio,* on occasions, is *Opilio parietinus* which has no dark, median band, but is a mottled greyish-brown with a row of longitudinal dots. It has spotted legs and a conspicuous elongated spot on the underside of the base of each leg.

Lepidoptera: Butterflies and Moths

Introduction to Chapters 8 and 9

In several ways, this order of insects is one of the most extraordinary. The two pairs of wings are expanded and are often very colourful, possessing a distinctive pattern which is sometimes highly complex. One usually does not have to be an expert to recognize the various species, and often the sexes can be easily differentiated. The colouring and patterning are brought about by a covering of over-lapping scales which, to the naked eye, appears as an easily-rubbed off, fine powder. Under magnification the powder is seen to consist of variously-shaped, plate-like elements. The latter are seen to possess a short stalk which pegs them to the wing membrane. The mouthparts are modified to form a long, coiled tube — the proboscis — for sucking liquids such as nectar from flowers.

There are three immature stages in their development. The egg hatches into a larva or caterpillar. This is elongated and segmented with each of the first three segments behind the head, furnished with a pair of jointed legs. There are additional 'false' legs on a number of the succeeding segments, including the last, and these are known as the anal claspers and are often larger. An exception is the large family GEOMETRIDAE in which the larvae are without the false legs and progress by alternately holding the substrate by the thoracic legs and the anal claspers. Almost all lepidopterous larvae feed on vegetable material — leaves, flowers or fruits of growing plants, and a small number on the woody tissues.

The active, larval stage is followed by the resting pupa or chrysalis which is often protected from predation in an enveloping cocoon or in some other way. For instance, they may be enclosed in a curled leaf, between two leaves, or wholly or partially underground in a small chamber. When the adult finally emerges the wings are small and crumpled, but a fluid is pumped through the 'vein' system, extending the wings to their full size. The distinction between Butterflies and Moths was, in earlier classifications, quite abrupt. Butterflies were placed in the sub-order *Rhopalocera,* in which the antennae were clubbed at the tip, whilst the Moths were *Heterocera,* in which the antennae were of all shapes. Today the distinction is more blurred. The Moths were again divided into two groups, quite artificially, according to size — small moths in the *Micro-Lepidoptera* (although no butterflies were included) and the large moths with all the butterflies in the *Macro-Lepidoptera.* In this chapter we describe only the Butterflies, leaving the Moths for the succeeding chapter.

A great debt is owed to E.S.A. Baynes for the years of painstaking work he put into cataloguing the species of Butterflies and Moths found in Ireland. According to Baynes' Lists and the Distribution Maps in South's *British Butterflies* (1973), there are 28 'permanently resident' butterfly species although some of these are very local in their distribution. The number is made up from 8 'Browns' or Satyrs; 6 Fritillaries and Vanessids; 3 'Blues'; 1 Copper; 3 Hairstreaks; 6 Whites or Pierids and 1 Skipper. In addition to these there are three summer migrants which, although varying greatly in numbers, occur in most years. They are the Red Admiral, the Painted Lady and the Clouded Yellow. Another six species are found as migrants only on very rare occasions. These are the Milkweed or Monarch, *Danaus plexippus*; Queen of Spain fritillary, *Argynnis (Issoria) lathonia*; American painted lady, *Vanessa virginiensis*; Camberwell beauty, *Nymphalis antiopa*; Bath white, *Pontia daplidice*; and the Pale-clouded yellow, *Colias hyale.* The status of the additional species has not yet been resolved. It is probable that they are extinct. These are the Small mountain ringlet, *Erebia epiphron,* and the Heath fritillary, *Mellicta athalia.*

8 Butterflies

Dingy Skipper *Erynnis tages*

One might be forgiven for thinking that this somewhat inelegant name for a butterfly denotes a rather drab species. True, the coloration of its wings is in browns and brownish-greys — rather tweedy in fact, but it is nonetheless beautiful for all that. The Dingy Skipper is a butterfly of the first really warm days of summer, which are usually in May but some years can be around the middle of April or could be as late as June. to many who do not know their Irish butterflies, it can be passed as a moth when resting on a flower or grasshead. When resting thus, it folds its wings downwards in a tent-like fashion and its club-shaped antennae are prominent. It is a butterfly of the dry, open places along old railway lines and glades deep in forests wherever the food plant of the larvae is to be found. This is Bird's-foot trefoil, *Lotus corniculatus,* and the Dingy Skipper favours no other plant for nurturing its caterpillars. These are yellowish-green and much stouter in the middle than at the head and tail, and the whole body is covered with minute, white spots.

Except for the Burren in Co. Clare, it is nowhere very abundant but modern records show it to be found in Sligo, Mayo, Galway, Tipperary, Kildare and Dublin, whilst in the south-east it still occurs in Co. Wexford. The Burren specimens are referred to the sub-species *baynesi* where the brownish ground colour is very much darker and the markings much lighter, making a much more contrasting pattern. The Dingy Skipper is the only species of the world-wide HESPERIDAE family to occur in Ireland.

Wood White
Leptidea sinapis ssp.juvernica

This fragile-looking butterfly is, nevertheless, a determined flyer and is well able to cover a lot of ground, so that it can (and does), recolonize old localities when the climatic and other factors are suitable. It is locally abundant mostly in west and east central areas.

It is creamish-white in colour with a blackish-grey tip to the forewings — conspicuous in the male, but not so in the female. The length across the wings is about 40mm. It is usually single-brooded, but is said to be double-brooded in the Burren area of Co. Clare. It first appears in May and June, and, if a second brood emerges, this will take place in July and August. The Irish Wood White, subspecies *juvernica,* is more strongly marked than the British specimens, and the hindwings are suffused with a yellowish-grey. The caterpillar is green, dotted with black at the front and with a dark-green line bordered with yellowish-green. It feeds on a number of

above: Dingy Skipper, *Erynnis tages*. Length across outstretched wings 29 mm. Abundant in the Burren, Co. Clare but occurs also in Counties Sligo, Mayo, Galway, Tipperary, Kildare, Dublin and Wexford.

above: Irish Wood White, *Leptidea sinapis juvernica*. Length across outstretched wings 40 mm. Fairly well distributed and locally abundant.

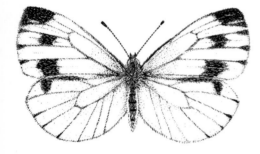

above: Green-veined White, *Pieris napi*. Length across outstretched wings about 42 mm. Generally distributed and the most plentiful of the 'Whites'.

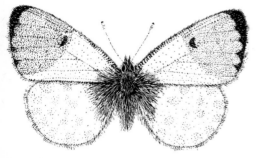

above: Orange-tip, *Euchloë cardamines hibernica*. About 40 mm across outstretched wings but sometimes only about 32 mm. Plentiful and generally distributed throughout Ireland.

species of LEGUMINOSAE, chief among which are Tuberous pea, *Lathyrus tuberosus*, Tufted vetch, *Vicia cracca*, and Bird's-foot trefoil, *Lotus corniculatus*.

Green-veined White *Pieris napi*

Six species in the family PIERIDAE occur in Ireland and four of them, the Large White, *Pieris brassicae*; the Small White, *Pieris rapae*; the Green-veined White, *Pieris napi*, and the Orange-tip, *Euchloe cardamines*, are generally abundant. Members of this family usually have a white ground colour, so are seen on the wing, even by the least observant person. The Green-veined White is generally distributed throughout the country and is the most plentiful of the 'Whites'. Being double-brooded, may be seen sometimes as early as late April — then throughout May and June — and again in late July, and throughout August.

The colour of the butterfly is mostly white, with light-grey veins. There are single spots on the forewings in the male, and two spots in the female. The apex of the forewings is marked in dark grey. The upper surface of the wings could be confused with the Small White, but the underside of the hindwings is bright yellow; with the veins made conspicuous with black scales, thus giving the appearance of green. The two broods may be distinguished by the intensity of the vein marking. The length across the outstretched wings is about 42mm.

There is much variation, however, in this species. Sometimes the spots on the forewings are enlarged, often running together, and there are spots on the hind wings too. This form is known as *ab.fasciata*. Occasionally the females are bright yellow on the upper surface, and are known as *ab.hibernica*. For some years these special Irish forms of the Green-veined White were bred artificially by butterfly collectors and dealers. The caterpillar is green with white and black hairs which arise from black warts.

Orange-Tip
Euchloë cardarmines ssp.hibernica
This butterfly is plentiful and is generally distributed throughout Ireland, where damp areas

support its main foodplant — Lady's Smock, *Cardamine pratensis* as well as the other related cruciferous species. A number of the latter are hedgerow plants or found in gardens; to some extent, it is associated with the activities of man. The colour of the butterfly is predominantly white, but in the male there is a large, orange patch on the outer edge of the forewings. The underside of the hind-wings of both sexes are patterned reticulately with green patches, this reticulation serving to disguise the insect when it is at rest on the green and white flowers of many of the hedgerow plants.

The Orange-tip appears in May and June, and the caterpillars feed during the end of June and July on the plants referred to above. They are bluish-green and rather hairy with a white band along the sides. The pupa is strangely elongated and sharply-pointed at the head end, with expanded wing-cases. It is held in position on a stalk by hooks at the base which hold on to a pad of silk, with a silken girdle in the middle. The Orange-tip of Ireland is a distinct sub-species, *hibernica,* differing from the British sub-species in being smaller; the dark markings at the end of the veins being more pronounced and, whilst the female has yellowish hindwings, the male has yellowish undersides to the fore-wings.

Green Hairstreak *Callophrys rubi*

Three Hairstreak species occur in Ireland — the Green, *Callophrys rubi*; the Brown, *Thecla betulae,* and the Purple, *T.quercus.* The Green is distributed throughout Ireland, and although it is said to be never very abundant, it was most pleasurable to see this agile, little butterfly in some numbers in May and June 1977 in a front garden at Dromreagh, Co. Cork.

The upper-sides of the wings are brown with faint, gold shading but the undersides are almost a brilliant, metallic green with a few, small white spots. It measures about 27mm across the outstretched wings. Its little, green, hump-backed caterpillars feed on a wide variety of plants — principal amongst them being gorse, *Ulex europaeus* — so that in Ireland it may never fear for a lack of its foodplant! The Brown Hairstreak today is extremely local and rare,

Green Hairstreak, *Callophyrs rubi.* It measures about 27 mm across the outstretched wings. Generally distributed, although fairly local, and sometimes abundant.

having been found (since 1963) only in three areas of Galway and Clare, but in former years it was recorded from Kerry, Cork and Wexford. This species measures about 42mm across the wings, the uppersides of which are dark-brown, with a large orange patch on the forewings in the female, but with only a small patch of dark orange shading in the male. The undersides of the wings are a light, orange-brown, streaked with black and edged with white.

The Purple Hairstreak is about 32mm across the outstretched wings, and in the male the fore-wings bear two patches of bluish-purple near the base; whereas in the female these patches occupy almost all the wing area with the ground colour black in both sexes. The larvae feed on oak, *Quercus robor,* and I also expect *Q.sessiliflora* forms part of their diet. This species seems to have disappeared from a number of localities where it formerly occurred and, since 1961, is known only from one area in the central-south and two areas in the central-east.

above: Peacock, *Nymphalis io*. About 61 mm across outstretched wings. Generally distributed but commoner in Southern half of country.

above: Silver-washed Fritillary, *Argynnis paphia*. About 62 mm across outstretched wings. Abundant in wooded areas in a number of widely scattered localities throughout the country.

Peacock *Nymphalis io*

The Peacock butterfly hardly requires further description. With a dramatic 'peacock eye' of blue and black on each wing, accentuated with cream on the forewings and grey on the hind-wings, set on a background of deep crimson. It *cannot* be mistaken for any other species. In marked contrast to the upperside, the underside is black in general appearance but close up is intricately lined with black, brown and grey. It has a general distribution in Ireland but is more common in the southern half than in the north.

Freshly emerged butterflies are seen in August and September, usually on sunny days where they are especially attracted to the spikes of Buddleia blossom. They then hibernate until the following spring – in hollow tree trunks and other similar situations, yet not usually found in buildings. The butterflies emerge during the first warm days of spring, and the batches of eggs are laid on the Stinging nettle *Urtica dioica*. The larvae feed in colonies during fine weather in June and July, and are black with small white spots, together with black spines. The variably coloured pupa hangs suspended from a silk pad by means of hooks on the tip of the abdomen. The author holds the Peacock in special regard, as it was the very first butterfly species he reared from the caterpillar stage after the common Whites.

Small Tortoiseshell *Aglais urticae*

Although having a rather similar distribution to the Peacock, the Small Tortoiseshell appears now to have become rarer in the north of the country. In some areas, however, it is especially abundant, such as in West Cork. This species is easily identified. The background is an orange red. Along the front margins of the forewings there are alternate rectangular patches of black and light yellow, with a white patch near the tip. The side-borders are black with a series of inset, blue spots. The undersides are patterned in black, biscuit and bluish-grey, so that when the insect is at rest, with wings closed, it resembles a piece of dead leaf; this often enables it to remain un-detected by predators.

This butterfly is double-brooded, the first brood appearing in June and the second in August and September. Those of the second brood hibernate almost as soon as they emerge, selecting old sheds and buildings in which to over-winter. Often they will hide away amongst curtains in dwellings and, when conditions are warm, the butterflies become restless, flying about the room seeking to make their exit. They should *not* be put outside, but into a cooler part of the house from whence they can escape when spring arrives. The caterpillars feed on Stinging nettle, *Urtica dioica,* feeding together in colonies amongst a loose web until shortly before pupation, when they wander away.

Silver-washed Fritillary
Argynnis paphia

This fine butterfly is abundant in wooded areas in a number of widely-scattered localities generally, near the coast. It measures about 62mm across the wings, although the male specimen illustrated was only 57mm. The upper-sides of the wings are bright, fulvous orange, easily seen at a considerable distance. There is also a chequering of black veins and spots, and

above: Marsh Fritillary, *Euphydryas aurinia*. Male about 34 mm, female about 40 mm across the outstretched wings. Widely distributed but local. Appears to have disappeared from many of its localities in recent years.

the female is rather paler in colour. It is from the undersides of the hind wings, however, that this butterfly derives its name. The undersides are pale greyish green and traversed by four rather wavy bands of silver with two bands of green marks along the outer edge.

These butterflies appear in July, but worn and ragged specimens are sometimes encountered in early August. The eggs are laid in the crevices of the trunks of trees — I know them only from oak, but other tree species have been recorded. Hatching in August, the minute larvae commence to hibernate immediately. In spring, they descend from the tree and feed entirely on the leaves of the Dog violet, *Viola canina*. The pupa hangs downwards from a tree trunk or other nearby object and it is able to wriggle actively and this is when conspicuous, gold spots are displayed.

The Marsh Fritillary
Euphydryas aurinia ssp.hibernica

This attractive fritillary butterfly is widely distributed in Ireland, although very local, only occurring in marshy areas where the larval food plant Devil's Bit scabious, *scabiosa succisa*, grows. It is a variable species wherever it occurs and in Ireland this appears to be especially true. The two forms of the butterfly to be found in Ireland are *ab.scotica* and *ab.praeclara*, the former being said to be more abundant than the latter. In *praeclara* the transverse band is straw-coloured, and the reddish areas are more vivid. The base of the wings is blacker, as are the veins and cross-lines. In *scotica* the black is even more intense, but the straw-coloured markings are dull.

The butterfly is about during May and June when the eggs are laid in batches on the leaves of the scabious. In June and July they hatch, and, after feeding for a time, the caterpillars construct a silken web late in August. They then retire into this for hibernation, and at the beginning of March they recommence feeding. When full-grown the caterpillar is black but has a number of white spots, each of which bears a short, black hair. The first thoracic segment is hairy whilst the second and third each bear two spines. The abdominal segments bear short, black spines arranged in nine rows. The head is

above: Speckled Wood, *Pararge aegeria*. Up to 44 mm across the wings. Widely spread throughout the country in woodland areas.

black, as are the true legs, but the ventral surface of the body, as well as the false legs, are dull, rust-red. In captivity the larvae will eat the garden varieties of scabious in addition to honeysuckle, *Lonicera periclymenum*.

Kane, in his *Catalogue of the Lepidoptera of Ireland*, instances extraordinary swarms of the Marsh fritillary larvae which have occurred in the past. At Ennis in County Clare, there was once said to be a 'shower of worms' which were identified as the larvae of this species which were migrating in search of scabious, and 'the black layer of insects seemed to roll in corrugations as they swarmed over each other'. The larvae appear to be heavily parasitized by Hymenoptera Parasitica, *Apanteles* being specially important in this respect.

Speckled Wood *Pararge aegeria*

This agile, medium-sized butterfly is widely spread throughout Ireland, and is quite abundant in suitable localities in the Burren area (County Clare), the far south-west of Cork and Kerry,

Galway and the south-east coastal areas. Elsewhere its distribution is very scattered, although it may be that when more detailed surveys are carried out the Speckled Wood may be found to be more extensively distributed than is at present thought.

The ground colour is a rich, very dark brown and the spots or 'blotches' are yellowish-cream. There is a black eye-spot near the tip of the forewings and three similar 'eyes' towards the edge at the rear of the hind wings. The female is slightly larger than the male, with wings of a rather rounded appearance. This butterfly is nearly always found in shady situations along the edges of woods, or lanes. The male appears to be territorial, as it will give battle to any other approaching male. The larva feeds on a number of grass species.

Gatekeeper Butterfly *Pyronia tithonus*
The central areas of the wings are rich, brownish-orange which are margined with brownish-grey.

above: Gatekeeper, *Pyronia tithonus*. Female about 44 mm across the wings and male a little smaller. Mainly in the South-west and South-east. Absent from the Northern half of the country.

Near the tip of the forewings is an eye-like mark consisting of a black spot enclosing two white dots. The female Gatekeeper is about 44mm across the wings, whilst the male is smaller, with a greyish-brown band across the central area of the forewings, and the white dots in the eye-mark may be reduced. The larvae feed on various grasses, and the adults fly during July and August. Irish and British examples have been given the sub-specific rank of *britanniae*. Distribution is mainly in the far south-west and in a few, scattered coastal localities in the south-east.

Meadow Brown
Maniola jurtina spp.iernes
We have eight species in the family SATYRIDAE in Ireland, making it our largest butterfly family. They are often popularly called "Browns" on account of their generally sombre colouration, although the Wall butterfly, *Lasiommata megera*, could not possibly earn this title. I prefer to call them the 'Satyrs' — after the woodland gods.

The Meadow Brown is very well distributed and generally abundant wherever it occurs. Many gaps in the Distribution Maps probably show the absence of recording entomologists rather than absence of butterflies! Its fuscous- brown ground colour has a suffusion of orange occupying most of basal part of the wing especially in the female, which also carries a broad, orange triangular band across the forewings. The male shows much less orange but both sexes possess a black eye-spot with a white pupil. The eye-spot is narrowly ringed with orange in the male of the species. In both sexes the undersides are much lighter in colour. Irish specimens are referred to the subspecies *iernes* which are larger and brighter in colour than those occurring in Britain.

The caterpillars are green — longitudinally striped with a darker green with white, and covered with short, white hairs. They feed on various grasses in the genus *Poa*. The adult butterflies may be seen on the wing at almost any time from June to September, although it is thought that second brood butterflies occur only occasionally.

above: Meadow Brown, *Maniola jurtina iernes*. Female about 50 mm across outstretched wings, male a little less.

Small Heath Butterfly
Coenonympha pamphilus
This little butterfly is the smallest of our Satyrs.

101

Small Heath,
Coenonympha pamphilus.
About 30 mm across
the outstretched wings.
Widely distributed.

In the resting position, as shown in the illustration, it measures only 20mm from the eye to the tip of the forewing. The uppersides of the wings are pale sandy brown and there is a narrow greyish-brown border. On the undersides the forewings are tawny, with a light-grey area at the apex, and with a small, but conspicuous, black, eye-like marking with a white pupil. The hindwings are of different tones of grey and rather hairy with a few small, white, eye-like marks which may be very faint or absent altogether. It is said to be widely distributed and common throughout Ireland, although, because of the lack of recorders, there appear to be a series of wide gaps in the Distribution Maps.

The Small Heath may be seen from May to August and it is partly double-brooded, some passing through the complete life-cycle from May to August, whilst others take a year. The green caterpillar has a dark stripe down the back and a lighter stripe on the sides. It feeds on Annual meadow grass, *Poa annua*; Wood meadow grass, *Poa nemoralis,* and Meadow fescue grass, *Festuca pratensis.* The closely-related Large Heath *Coenonympha tullia,* is abundant in a number of extensive boggy regions, mainly in the centre and the west of the country. It is an extremely variable insect and Irish specimens are usually referred to the subspecies *scotica* and *polydama.* Whereas in most of the subspecies there are rows of prominent eye-spots, on the underside, eye-spots are almost absent in *scotica.*

9 Moths

MOTHS *Lepidoptera-Heterocera*

The attention of the reader has already been drawn to the new classification of the *Lepidoptera*, where the distinctions between the butterflies and moths are not so finely drawn as they had been hitherto. In fact no problem exists as to which species of Lepidoptera are butterflies and which are moths, and this chapter describes and illustrates a number of moth species. Examples of all the families found in the nearest parts of continental Europe are present in Ireland.

Like the butterflies, moths possess large, ornamental wings, the patterns and colours of which may be differentiated with ease. Indeed, they often exhibit much more besides their specific determination. Some species, often on account of some degree of isolation from the general population, show more or less stable genetic differences, and at least they can be recognised and their geographic location established. Again, so far as recording the presence and distribution of moths in Ireland, we are tremendously indebted to the work of E.S.A. Baynes. He lists no less than 538 Irish moths, so that what follows can only be an arbitrary selection.

Death's-Head Hawk-Moth

Acherontia atropus

Twenty-seven species of SPHINGIDAE or Hawk-

Death's Head Hawk. *Acherontia atropus*. About 10 mm across the wings. This migrant recorded most years.

moths occur in Europe, and the Death's-head is the largest of them all. In Ireland it occurs only as a migrant as it seems not to be able to maintain itself over the winter. Nevertheless, it is recorded in most years from some part of Ireland. There were ten reported in 1956.

The adult moth is about 10cm across the outstretched wings, and whilst this expanse is attained by other Hawk-moths, the Death's-head possesses a very thick body and the wings are also wider. Its forewings are greyish-brown and reddish-brown and the hindwings are yellow with two greyish-black bands, the outer band broader than the inner, but the most noticeable characteristic is the 'skull and crossbones' pattern on the thorax, from which the species derives its popular name.

The full-grown caterpillar is about 12cm in length, and although usually some shade of green, is sometimes yellow, brown or almost black. Its body is covered with white or violet tubercles and the seven oblique stripes, associated with the Hawk-moth larvae, are purplish, edged with yellow. There is some variation, however, as in the general body colour, and the horn has a double curve. The larvae generally feed on the leaves of potato, 'Tea tree', *Lycium barbarum*; Woody nightshade, *Solanum dulcamara*; and Snowberry, *Symphoricarpus spp.* The larvae are more often found than are the adult moths, as they are observed when the potatoes are harvested. The larva pupates 5 to 10cm in the ground in a rather flimsy cocoon.

The adults are normally found in May and June, but there is a controversy as to whether these are migrants or have naturally emerged from over-wintering pupae. The proboscis is short and unadapted to taking nectar from the corollas of flowers, as in the case of other Hawk-moths, but in the days when bee-hives were straw skeps the Death's Head was often known to have robbed the bees of honey. In the past it was known as the 'Bee-robber' and 'Bee Tiger'. When handled the moth emits a shrill squeak! Abroad the species is distributed widely, being found in central and southern Europe, North and South Africa and Asia generally.

Eyed Hawk-Moth *Smerinthus ocellata*
A large moth, the Eyed Hawk is said to be widely distributed throughout Ireland and to be moderately common, especially so in boggy areas. The food plants of its larvae are willow, and sallow, *Salix spp.,* apple, *Malus domestica*; and crab apple, *Malus sylvestris,* but poplar and privet have also been recorded. The eggs are usually laid not more than two on a leaf. The caterpillar is green (which may be yellowish or greyish), and covered with white tubercles which give a rough texture, and there are seven whitish, oblique stripes on the sides. These are edged with green at the front. A violet hue is sometimes present. Its bluish horn is green towards the tip, which in turn is dark in colour. The triangular head is also bluish.

Eyed Hawk, *Smerinthus ocellata*. 8 cm across outstretched wings. Widely distributed and moderately common.

The forewings of the adult moth are patterned in various tones of brown, but the hind wings are suffused with carmine, and on each wing there is an extraordinary life-like 'eye' marking in light and dark blue. These markings only become visible when the wings are held forwards. The outstretched wings measure 8cm across. The moth then possesses two progressive stages of protection; firstly it merges into its surroundings and secondly, if this has failed and it has been detected by a predator, it can flick its wings forwards and frighten its assailant by the large pair of 'eyes'.

Poplar Hawk-Moth *Laothoe populi*
Another large moth, up to 9cm across the outstretched wings, the Poplar Hawk is widely distributed and abundant throughout Ireland. It is to be found also on some islands where the food plants of poplar, *Populus serotina*, aspen, *Populus tremula,* willow and sallow occur. The forewings and the outer part of the hindwings are grey and brown with some variation of intensity of colouration, but the base of the hindwings possesses a conspicuous brick-red patch. One would not have thought that this was sufficient to frighten off a would-be predator, as its close relative, the Eyed Hawk moth, might have done.

The eggs are relatively large, like shining, green pearls and laid in May, usually singly on the leaves of the food plant. When fully-grown, the caterpillar is green, with a rough appearance given by yellowish tubercles. The seven oblique

Poplar Hawk, *Laothoe populi*. Up to 9 cm across the wings. Widely distributed and abundant.

White Prominent, *Leucodonta bicoloria*. Measures about 35 mm across the wings. First found in Killarney in 1858 but has not been seen in Ireland for over forty years.

stripes are yellow, the spiracles are reddish and the horn is green, sometimes tipped with a reddish colour also. Although the caterpillar head is triangular, it is not so sharply pointed at the top as is that of the Eyed Hawk-moth. Two broods often occur in the year, and three have been recorded in exceptional circumstances.

The White Prominent
Leucodonta bicoloria

The White Prominent is classified in the NOTODONTIDAE, a family characterized by the possession of a tooth-like tuft of scales projecting from the middle of the inner margin of the forewings. When the moth is at rest the tufts come together and project above the closed wings. The males have bipectinate or toothed antennae. The larvae of a number of the 'Prominent' group have one or more pronounced humps on the back, and rest in a peculiar posture. The hind end, and sometimes the front, is elevated, so that only the abdominal prolegs (but not the anal pair) clasp the twig on which it is standing.

The White Prominent is not only the rarest of the twenty-five or so species of the NOTO-DONTIDAE found in Ireland and Britain, but is also one of the most local and rarest of *all* moths. It was first found in Killarney in 1858, but unhappily the last to be seen in Ireland was over forty years ago. This may be due to its retiring habits, although its bright colouration and the use of the outstanding attractant powers of the Mercury Vapour lamp (used for

catching moths), lead one to believe that the Killarney population has become extinct. The ground colour is a glossy white and a three-pronged orange and black mark occupies the centre of each forewing. It measures about 35mm across the outstretched wings.

The pale, yellowish-green caterpillar has a whitish upper surface with a number of longitudinal green lines, the central one of which is darkest. On each side of the caterpillar there is a yellow stripe edged with green which touches the spiracles at their base. It feeds on the leaves of silver birch and becomes fully fed in July, when it pupates within a strong, silken cocoon spun amongst the leaves. It emerges the following May or June.

The Poplar Lutestring
Tethea or ssp.hibernica

This attractive member of the family THYA-TIRIDAE has pearly-grey forewings tinged with pink near the base, and there is a dark band with a jagged outline made up of four black lines (lutestrings). Near the outer margin there is a paler band made up of a number of lines of differing intensity. It measures about 35mm across the outstretched wings. This Irish sub-species is found in Ireland locally where aspen occurs in Donegal, Fermanagh, Sligo, Cavan, Mayo, Galway, Wicklow and Kerry.

The larvae are yellowish-green with a dark line along the back, and two black spots behind the yellowish-brown head. It generally hides during the day time between a pair of leaves which have been spun together by the caterpillar. The moth is found during June and July.

Chinese Character *Cilix glaucata*
The white forewings bear a grey-brown saddle-shaped mark. Silvery scales are present on the veins. It is about 20mm across the outstretched wings. Recorded mostly from the northern half of Ireland, it has however been found in Clare and Cork.

The caterpillar feeds on hawthorn, *Crataegus monogyna,* and sloe, *Prunus spinosa,* and more rarely on apple and pear, and the double-brooded adult emerges in May, June, and late July and August. The adult moth has the curious habit of sitting openly on a leaf — like a bird-dropping — and if disturbed, drops to the ground shamming dead.

The Drinker *Philudoria potatoria*
Fairly large, robust and furry, the Drinker moth appears to be widely distributed throughout Ireland, although only in a few locations is it abundant. Even though the adults are attracted to light, they are much less in evidence than are the large, hairy larvae (caterpillars) which sit out on grass stems on which they feed, on road verges in rainy weather — hence the name 'Drinker'. The male moth is reddish-brown with two small silvery marks in the centre of the forewings which also carry an oblique, transverse bar stretching across them. The antennae are extremely feathery. The female moth is larger than the male, and altogether much lighter in colour. The adults emerge in July.

The Dew Moth *Setina irrorella*
The Dew moth is classified with the 'footmen' in the sub-family LITHOSIINAE of the ARCTIIDAE. In colour it varies from a yellowish-buff to creamy-white, with the margins somewhat deeper. The more yellowish forewings bear two rows of black spots transversely, with a few along the outer margin as well as several near the outer margin of the hind wings. There is some variation in the numbers and colour intensity of the spots. The thorax and abdomen are black except for tufts of yellow hairs on the former which also occur at the tip of the abdomen. In size it is about 27mm across the outstretched wings.

The larvae feed on lichens found on rocky hillsides and sea-cliffs. The adults emerge in June and July. In Britain the distribution is scattered from the coasts of Kent and Sussex, Isle of Wight, St. David's in Wales, and in Scotland in Aberdeenshire, Sutherland, Tweed, Tay, Clyde and Argyle. In Ireland it is local, but where it occurs it is exceedingly abundant — such as in south Galway, in the Burren, County Clare, and in the Aran Islands. The larvae appears in vast swarms on the rocks after rain.

The Muslin Moth
Cycnia mendica ssp.rustica
This is a common moth in Ireland, being widely distributed from Antrim to Cork. It is classified with the 'Tigers' and 'Ermines' in the sub-family ARCTIINAE of the ARCTIIDAE. The females are white with a few, scattered blackish spots. The male, however, is dark, greyish-brown — very different from the female. Many extreme varieties are known, from almost complete suppression of the spots, to specimens heavily clouded with grey or rayed with black. In the Irish sub-species, named *rustica,* the males resemble the females in colour, although a few may be creamish or greyish-brown. The thorax and abdomen are black but densely covered with long, white hairs. The male illustrated measured 34mm across the outstretched wings.

The caterpillar has a greyish-brown ground colour but is covered with yellowish-brown hairs which arise from pale brown warts. There is a pale line running along the middle of the black, and a white, broken line on each side below the black, outlined spiracles. The glossy head is pale, chestunut brown. It feeds on many herbaceous plants including dandelion, *Taraxacum sp.,* dock, *Rumex sp.,* plantain, *Plantago sp.,* chickweed, *Stellaria media,* but it will also eat the leaves of birch, *Betula pubescens,* and rose, *Rosa sp.* A silken cocoon with which is mixed the caterpillar hairs and are camouflaged with earth. The moth emerges during May and June.

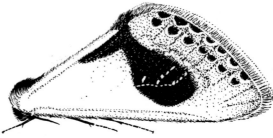

ove: Poplar Lutestring, *Tethea or hibernica.* Measures
out 35 mm across the wings. Found locally where
pen occurs in Counties Donegal, Fermanagh, Sligo,
van, Mayo, Galway, Wicklow and Kerry.

above: Chinese Character, *Cilix glaucata.* About 20 mm
across the wings. Mostly found in the Northern half of
Ireland but has been recorded in Counties Clare and Cork.

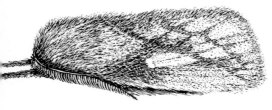

ove: The Drinker, *Philudoria potatoria.* Up to 62 mm
ross the wings. Widely distributed but abundant only
a few localities.

above: The Drinker, *Philudoria potatoria.* The large,
hairy caterpillar is more often observed than the
adult moth.

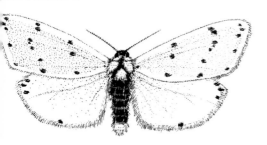

ove: Dew Moth, *Setina irrorella.* About 27 mm
ross the wings. Local but often extremely abundant.
ounties Galway, Clare and Arran Islands.

above: Muslin, *Cycnia mendica rustica.* About 34 mm
across the wings. Widely distributed and not uncommon.

above: The Grey, *Hadena caesia mananii.* Measures
32 mm across wings. Not uncommon in a number of
coastal localities.

above: Podlover, *Hadena lepida capsophila.* Measures
about 32 mm across the wings. Abundant in many
coastal localities.

The Grey *Hadena caesia ssp.mananii*

This Noctuiid moth of a most unusual colour is classified within the sub-family HADENINAE. As far as is known to the writer, this species is not found in Britain, except for the Isle of Man, but the sub-species *mananii* is local but not uncommon where it occurs in a number of maritime localities around Ireland, from Donegal to Clare, Kerry, Cork and Waterford. In addition, there are records from Co. Tyrone. On Inishvickillane in the Blaskets, a darkish-blue form occurs and this is exceptionally attractive.

It measures about 32mm across the outstretched wings. The forewings are slate-grey to a dark bluish-grey — a colour seldom found in nature. The basal third of the wings is darker and there is a lunate mark about halfway along the front margin. The hindwings are dark silver-grey, darker towards the outer margin, but there is a whitish fringe. The caterpillar feeds on the seed heads of the campions, *Silene maritima,* and *S.inflata.* It is pale ochreous brown in ground colour but is minutely spotted with a darker tone, and there are a number of blackish but indistinct lines along the back. The pale brown head is marked with a darker colour. The moth may be found during June, July and August, and generally frequents the flowers of the larval food plants.

The Pod Lover
Hadena lepida ssp. capsophila

The Pod Lover has the typical stout-bodied appearance of the NOCTUIDAE and the forewings have a very distinct pattern of greyish-white marks on a dark, brownish-grey background. It measures about 32mm across the outstretched wings. This is an abundant species in many localities around the Irish coast where the larvae feed on the seed-heads of *Silene maritima* and *S.inflata.* In the west and south-west the moths are dark in colour and have been named *ab.suffusa.* Many, however, are uniformly greyish-black and have been named *ab.obsolescens.*

The Marbled Green
Cryphia muralis ssp.westroppi

This extremely variable moth is classified in the sub-family ACRONYCTINAE of the fami NOCTUIDAE. It is about 25mm across the o stretched wings and is robust in appearance. England it ranges along the south coast fro Kent to Cornwall, and is also found in the Sci Isles. Elsewhere it is extremely local. It shou be looked for on lichen-covered rocks and o walls.

The ground colour varies from a light silve grey through greenish and bluish shades almost black. The attractive, dark grey mar ings may be distinct, through various inte mediate shades, to obscure. The outer margins the forewings are fringed with alternately whi and dark grey. The hindwings are silver-gre fringed with white. In Ireland a number extreme colour forms have been collected fro a few localities, of which Cork City has bee outstanding. D. Westropp brought these to th attention of entomologists. A form in colou resembling rich milk chocolate and found Cork City and Bandon, Co. Cork, has bee named *C.m. castanea.* Sub-species *C.m.simi* also occurs in Cork City and Bandon, Co. Cor Several extreme forms have been collected b H. C. Huggins at Dingle, Co. Kerry, one of whic *C.m.nigra* is black with white markings. Othe localities for this most variable moth ar Killarney, Co. Kerry, Clonmel, Co. Tipperar Dungarvan, Co. Waterford and Lough Erne Co. Fermanagh.

The caterpillar, which feeds on lichens, green with white, shining raised spots. The fir thoracic segment bears a black plate and thre broken, yellow lines run along the back. It ma be found amongst the lichens from October t May. It hides in a silken retreat during daytime The moth emerges during July and August, bu may be found, occasionally, in September.

The Sandhill Rustic
Luperina nickerlii ssp.knilli

In the past there has been a certain amount o confusion over the naming of this noctuiid mot of the sub-family AMPHIPYRINAE. The firs English specimens obtained from the Lancashire and North Wales coasts were named *Luperina gueneei,* but these are now referred to as a sub species, *gueneei* of *L.nickerlii.* The Irish sub-

above: Marbled Green, *Cryphia muralis westroppi*. About 25 mm across the wings. Most abundant around Cork city.

above: Sandhill Rustic, *Luperina nickerlii knilli*. Measures about 24 mm across the wings. Confined to the coast of the Dingle Peninsula and the Isle of Aran.

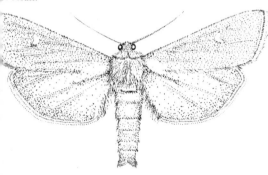

above: The Burren Green, *Calamia tridens occidentalis*. Only from the Burren, Co. Clare.

above: Brimstone, *Opisthograptis luteolata*. About mm across the wings. Occurs abundantly throughout eland.

above: Yellow Shell (dark form).

above: Yellow Shell (pale form), *Euphyia bilineata isolata* and *C.c. hibernica*. About 25 mm across the wings. Blasket Islands for the former and the Kerry Coast for the latter.

above: Argent and Sable, *Rheumaptera*. Measuring from 25-30 mm across the wings. Occurs mainly in the North-west, West and South in boggy areas.

above: Netted Pug, *Eupithecia venostata plumbea*. 18-24 mm across the wings. The subspecies is confined to the Blaskets and coastal localities of Counties Clare and Kerry.

species *knilli* is much darker in colour than the English sub-species *gueneei,* and it is confined to the coasts of the Dingle Peninsula in Kerry and the Aran Islands, where it occurs on the sandhills.

The forewings are sandy-brown in ground colour with some darker markings including a black X-shaped mark. There are some whitish spots and there is a kidney-shaped and an oval mark in the middle of the forewings which are also marked with white. The hindwings are shining white and margined with light brown moon-shaped marks. The moth measures about 24mm across the outstretched wings, and emerges in August. Little appears to be known of this Irish insect.

The Burren Green
Calamia tridens ssp.occidentalis
One of the most exciting entomological finds this century was the capture by W. S. Wright in the Burren, Co. Clare, of *Calamia tridens ssp. occidentalis* in 1949. It has been many years since a new moth species, of relatively large size, has been found within the British Isles. It created quite a stir amongst moth collectors and they flocked to the Burren. It was first thought to be the species *Luceria virens,* but subsequent research placed it amongst the group of moths in the family NOCTUIDAE — known as the 'wainscots'. But a species less like the drab wainscots would be hard to imagine. The forewings are bright, sage-green relieved only by the white kidney-shaped mark and a much smaller, oval mark in the centre of the wing. A fine, rust-red line edged with white defines the outer margin of the forewings. The head and thorax are also of the same sage-green. The hindwings are silvery-grey with a lighter margin.

The Yellow Shell *Euphyia bilineata*
ssp.isolata and *E.b.hibernica*
This is one of the most abundant of the family GEOMETRIDAE and is to be found generally distributed in the British Isles. The ground colour is usually yellowish-grey and a number of fine, brownish or greyish lines occur transversely across the wings, often forming two bands with

two angles on the outer band pointing outwards. This moth is smallish to medium in size, being about 25mm across the outstretched wings. The sub-species *E.b.isolata* is an extremely melanic blackish-brown form which was first found prior to 1893 on Tearaght Island of the Blaskets. It is now known to occur also on Inishvickillane, also of the Blaskets. On the Kerry coast and also opposite the Blaskets, and in West Cork, forms of this species have the wings a dark, fuscous brown. In addition there are specimens with dark, fuscous brown forewings and ochreous brown hindwings which are referred to as the sub-species *hibernica.* Another sub-species found in Ireland of this most variable moth is *testaceolata,* which is rather small and dull coloured and found along the Cork and Kerry coasts.

Generally the caterpillars are green to yellowish, stout with a central dark green line and yellowish lines on the sides which run along the back. In addition there is a pale, wavy line low down on each side, and the intersegmental grooves are also pale in colour. Some caterpillars, however, are pale, greyish-brown inclining to reddish-brown. They feed on many low-growing, herbaceous plants such as grass GRAMINEAE, dock and chickweed from August to May. The adult moths emerge throughout the summer.

Argent and Sable
Rheumaptera hastata
What enjoyment the old entomologists had naming our moths! It is 200 years since Harris named this black and white (but very variable) species. Across the outstretched wings it measures from 25-30mm. It is a day-flyer when sunny, and although comparatively small with a slender body its flight may be followed for long distances. In Ireland it occurs mainly in the north-west, west and south, and generally in boggy areas.

The caterpillar feeds on birch, *Betula, Vaccinum sp.,* chiefly, and also on sweetgale, *Myrica sp.* It lives in a shelter made by spinning the terminal leaves of the food plant together, and the adult emerges in May and June.

above: Speckled Yellow, *Pseudopanthera macularia*. About 29 mm across the wings. Locally abundant in the West (The Burren especially) and in the South.

above: Transparent Burnet, *Zygaena purpuralis sabulosa*. 25-31 mm across the wings. Locally in the Counties Galway and Clare.

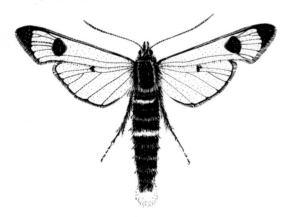

above: Welsh Clearwing, *Aegeria scoliaeformis*. Measures about 30 mm across the wings. Occurs in Killarney and Kenmare in Co. Kerry.

The Netted Pug
Eupithecia venosata ab.plumbea

Forty-two species in the genus *Eupithecia*, together with a further six in closely related genera, constitute the 'pug' section of the family GEOMETRIDAE. Forty-one of these occur in Ireland. To the lepidopterist, the pug has a characteristic shape (but difficult to define), but without exception they are small in size. Pugs are usually identified by their habit of resting with wings outstretched and held close to the tree-trunk, wall, etc., on which they are resting. The Netted Pug is a variable insect, but in its typical form is pale-greyish with black, transverse marks, two of which are edged with white. Several colour forms have been found in Ireland, but *E.v.plumbea*, named by H. C. Huggins, is leaden in colour with the transverse markings much broken. This abberant is confined to the coastal districts of Clare and Kerry, and in addition, the Blaskets.

The caterpillar feeds in the seed-capsules of catchfly, *Silene inflata* and *S.maritima*, and other species of *Silene*, from June to August. It is rather stout in appearance, greyish-brown above, but pale green or yellowish below. There are three dark brown lines along the back, and the head is blackish. The moth emerges in May and June and in its many forms is generally to be found wherever a food plant is abundant.

Brimstone Moth *Opisthograptis luteolata*

This delicately marked, lemon-yellow moth is abundant and occurs throughout Ireland. Measuring about 30mm across the wings, it has rusty-red blotches along the front margins of the fore-wings and there is a small, ring-like mark in the centre of the front margin together with ashy-grey marks running round two transverse bands on each wing.

The caterpillar resembles a twig, even to the extent of bud-like projections on segments six and eight. It feeds on hawthorn preferably, but occasionally on sloe and plum, *Prunus domestica*. Although the adult moth usually emerges during May and June, it has been recorded from April to August. The author writes with some nostalgia, as this was one of the first species he was able to identify, and each morning, over sixty years ago, he would race across to a hawthorn hedge in early summer to see how many had emerged, and in a knockabout world he would marvel at the delicate and immaculate scaling of this beautiful insect.

Speckled Yellow
Pseudopanthera macularia
A colourful little moth, the Speckled Yellow derives its name from the bright yellow wings being blotched and spotted with black. It is about 29mm across the wings. Although occurring in a number of counties in the west and south of Ireland, it is local in distribution but abundant where it occurs, being especially plentiful in the Burren, Co. Clare.

The green caterpillar bears white lines and feeds on woodsage, *Teucrium scordonia,* woundwort, *Stachys spp.,* and deadnettle, *Lamium spp.* The adult emerges in late May and in June, usually in open woodland. In warm, sunny weather it flies in daytime.

The Transparent Burnet
Zygaena purpuralis sabulosa
The Burnet and Forester moths are classified in the family ZYGAENIDAE. In Britain ten species occur whilst in Ireland only four are to be found. Of the latter, however, a sub-species of the Transparent burnet is of exceptional interest.

The body of the Burnet moths is characteristically of a bluish-green, metallic ground colour, whilst the forewings are metallic green with five or six carmine spots. The hindwings are of carmine but are margined with green. The antennae are clubbed, resembling those of butterflies. They are said to be extremely poisonous to birds, and certainly they rest and fly slowly in the sunshine quite openly, without hiding from birds and other predaceous animals that might prey on them. The metallic colouration is a warning signal that they are unpalatable.

The Transparent Burnet occurs in a number of forms throughout the British Isles where it is very local but, like all of this family, tends to be present in colonies, thus being abundant where they occur. This moth occurs, or *did* occur, at one locality in Wales and one in Scotland (at Oban and Loch Etive) and it has been recorded from Cornwall in England. Quite abundant colonies occur in Ireland — in Clare and Galway, wherever limestone is to be found. The sub-species is usually *hibernica,* but examples that resemble more closely the Scottish sub-species *caledonensis,* have been named *sabulosa.*

The caterpillar is dark green, more olive above but paler below. The outer row of spots is black whilst the inner row is yellow and there is a rather obscure whitish central line. It feeds on wild thyme, *Thymus praecox,* as well as burnet saxifrage, *Pimpinella saxifraga,* but other plants have also been recorded as being its food plant, including *Trifolium* and *Lotus.* The moth emerges in June. A word of warning — do not let your enthusiasm run away with you when searching for this insect, especially above the Cliffs of Moher!

The Welsh Clearwing
Aegeria scoliaeformis
This comparatively large member of the family SESIIDAE is easily identified by the two yellow bands around the abdomen, and by the large, chestnut-coloured tail tufts. It measures about 30mm across the outstretched wings. It takes its popular name from first having been recorded from Llangollen in North Wales, and it formerly had a scattered distribution in England — mainly in the midlands. Today it occurs in Killarney and Kenmare in Co. Kerry.

The larvae feed on the inner bark of large Birch trees and when fully grown (about May), construct a cocoon just beneath the surface of the outer bark. Emergence takes place in June and July. Many of the old birch trees carry a considerable population of this species. Baynes states that his Irish specimens were larger than those collected at Rannoch, Scotland.

10 Other Insects

Bugs, Dragonflies, Beetles, Flies

Maritime Bristletail
Machilis (Petrobius) maritima

The Maritime bristletail is to be found around almost the whole Irish coast wherever there are turfy, or rocky cliffs. It is confined, however, to a narrow strip of land just above high tide mark, and if a piece of loose rock or overhanging turf is removed they will be seen to give a quick jump and then to run for shelter. It is a member of one of the most primitive group of insects — the *Thysanura*, in which wings are not developed, but neither are they derived from winged groups. The torpedo-shaped body is about 12mm in length and the antennae are slightly longer than the body length. The eyes are situated dorsally, and a pair of jointed palps are bent backwards. In the tail region there is a long, central bristle, two shorter lateral bristles and a short ventral one. The whole body is covered with minute scales and the colour varies from light-brown to black and grey.

Northern Emerald Dragonfly
Somatochlora arctica

Dragonflies and damselflies constitute the insect order *Odonata*. They are characterized by having two pairs of long, narrow, net-veined wings, very large compound eyes and a very long, slender abdomen. Their powers of flight are extra-ordinary. They are predaceous — catching and eating flying insects whilst they are on the wing. The immature stage is known as a nymph, which much resembles the adult, except for the wings which grow externally at each instar. The nymph is entirely aquatic and breathes by means of tracheal gills located at the end of the abdomen, and are either appendages or are internal folds of the rectal wall. Nymphs are also predaceous, seizing small animals by means of the highly modified lower lip which is hinged and can be shot forwards to grip the prey. It is generally believed by country folk that they sting, but in fact they are harmless.

The *Odonata* are divided into two main groups — *Anisoptera* (Dragonflies) and *Zygoptera*, usually called damselflies. The former are larger and more robust than the latter, with unstalked wings. The latter, on the other hand, have a more delicate appearance and have stalked wings. The legs deserve mention because, in flight, the prey is grasped by all six of them and then moved to the mouth for consumption — the six claws meet at the mouth.

The method of mating is unique in the insect world. Before actual copulation the male must transfer his sperm from near the tip of his abdomen at the ninth segment, to the second segment, by bending his abdomen in a curve. The male grasps the female around neck, head or prothorax, according to species. The female then extends her abdomen downwards and forwards to reach the sperm with the end of her abdomen.

There are 22 species of *Odonata* in the Irish fauna, but some are extremely local or rare. The Northern Emerald dragonfly is one of the latter. Its wing span is about 68mm, and the length of the abdomen is 37 to 40mm. There is little size variation in the male and female. The eyes and prothorax are dull apple-green, and the abdomen is a dull, black-bronze colour, and in the female there is a pair of orange spots near the base of the third segment. The wings of the male are a pale yellow hue, but are almost clear in the female. The Northern Emerald is a rapid and erratic flyer, and is to be seen in its localities during the month of July and at the beginning of August. It breeds in peaty bogs and on the edges of muddy streams by pushing its eggs into the mud in the shallow water. It occurs (or occurred) only in Kerry and in western and central Scotland, but is absent from England. Its most recent records, however, were about 1946.

Irish Issus *Issus coleoptratus*
The family ISSIDAE in the *Homoptera* group is fairly large in numbers, having about 1,000 species worldwide. They are mostly dull in colour and the forewings (the tegmina) are often of a strange pattern and texture, and peculiar in shape, with raised, vein-like ridges. The Irish species *Issus coleoptratus* has a squat appearance. The tegmina are roughly triangular in shape, with greenish patches on occasion, and are held horizontally. The legs are relatively long and are held out in the same plane as the tegmina. The hind wings are of a smokey brown colour. Measuring about 6mm in length, the tegmina are almost as wide, when the insect is at rest, as the body is long.

Issus coleoptratus is usually found in holly and ivy, but may occasionally be located on trees of various species, and it has even been collected from moss. Found throughout Europe and North Africa, *I.coleoptratus* is on the whole considered to be a southern species. Nine other species are

to be found in Europe, but the latter is found in Ireland and not in Britain.

Irish Water Boatman *Sigara fallenoidea*
The insect order *Hemiptera* is said to be the dominant group of those insects which do not undergo a larval stage. There is a very wide spectrum of shape and size, but the unifying character is the shape of the mouthparts. These are adapted for piercing and extracting liquids. By far the larger group, the sub-order *Homoptera*, are parasitic on plants — sucking their juices. It is characteristic of this sub-order that the wings are membraneous, and are held tent-wise over the abdomen. Perhaps the most well-known example of this is the *Aphis* or greenfly. There are 39 families in this sub-order. In the second sub-order, the *Heteroptera* there are no less than 47 families. The most easily recognized character for their indentification is the possession of wing-cases of a specialized type, and these have a basal, thickened part known as the corium, as well as an apical membraneous part, and usually fold flat — scissor-like over the abdomen.

Several groups of families are aquatic and one group, containing the Water striders, GERRIDAE, live on the water surface, feeding on dead or living animals found there. Another group consisting of the family CORIXIDAE, or Water boatman, suck mostly algae and detritus from the bottom of the pond but are occasionally carnivorous. The Water boatman possesses a scoop-like tarsus on the front pair of legs — the palea, and these are used for gathering food on the pond-bottom. Its long, thin legs are used for clinging, whilst the fringed, hind legs are used for swimming. The Irish Water boatman has paleae of a very large size, and these can be distinguished with the naked eye. In addition, the fringe of hairs on the femur of the middle pair of legs does not quite reach the tip. The distribution is primarily in northern Europe, and has so far not been found in Britain, but occurs in western and central Ireland.

The Clathrate Ground Beetle
Carabus clathratus
The beetles or *Coleoptera*, are said to be the

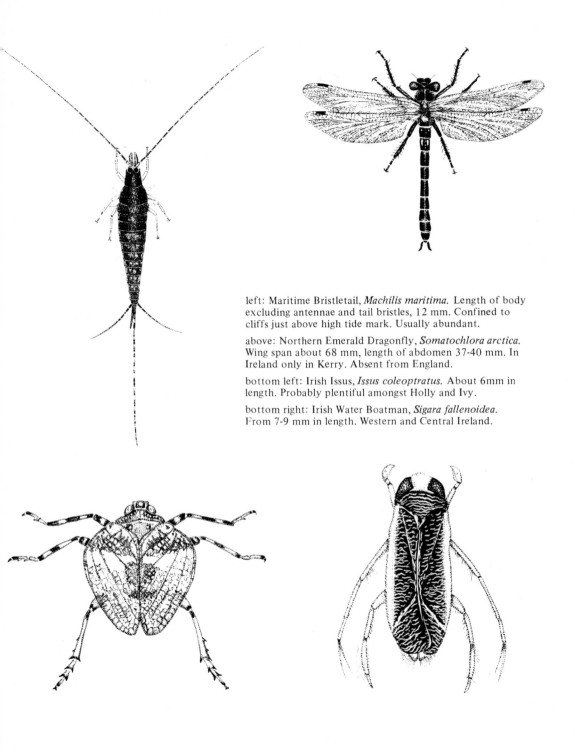

left: Maritime Bristletail, *Machilis maritima.* Length of body excluding antennae and tail bristles, 12 mm. Confined to cliffs just above high tide mark. Usually abundant.

above: Northern Emerald Dragonfly, *Somatochlora arctica.* Wing span about 68 mm, length of abdomen 37-40 mm. In Ireland only in Kerry. Absent from England.

bottom left: Irish Issus, *Issus coleoptratus.* About 6mm in length. Probably plentiful amongst Holly and Ivy.

bottom right: Irish Water Boatman, *Sigara fallenoidea.* From 7-9 mm in length. Western and Central Ireland.

most successful of the insect orders in having adapted themselves to the greatest variety of ecological situations. The number of beetle species in the world is in excess of a quarter of a million, indeed, almost three-quarters of the known animal species are beetles! The success of their numbers is thought to be associated with the conservation of their body-water. The principal characteristic of beetles is the modification of the first pair of wings into hard, horny wing-cases or 'elytra', which protect the membraneous second pair of wings which are folded flat under them. What is thought to be of the utmost significance, however, is the position of the breathing spiracles. In all other insects with the exception of beetles, the spiracles open directly to the external environment, whereas in the *Coleoptera* they open into the space between the wing-cases and the upper surface of the abdomen. This creates a moisture-gradient between the external environment and the water content of the insect itself. The largest and smallest insects are beetles!

The family CARABIDAE is one of the most primitive and unspecialized of the *Coleoptera,* and commonly finds a place at the beginning of books on beetles! They are predaceous — both in the larval and adult stages — and are quick running. The antennae are usually thread-like and of eleven segments, and the biting mouthparts are prominent. The prothorax has well-defined margins and the wing-cases are often fused along the middle line and the wings are often absent. The Clathrate ground beetle is said to be a northern species. Common in Scotland and Ireland, but rare in England, it is basically very dark in colour with a brassy or greenish sheen. Down each elytron there are three longitudinal rows of large, metallic pits or pores which are situated between ridges. It is a common species found locally around the edges of blanket bogs and along stream margins. The specimen illustrated was taken from a reclaimed blanket bog in Co. Mayo.

Eurynebria complanata

The appearance of this agile, predaceous member of the family CARABIDAE is quite distinctive. It is about 20mm in length and the pale straw

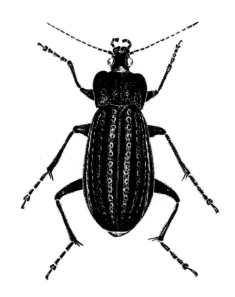

above: Clathrate Ground Beetle, *Carabus clathratus.* From 22-28 mm in length. Common.

below: *Eurynebria complanata.* About 20 mm in lengt Rare, found only in a few coastal localities in Counties Wicklow and Wexford.

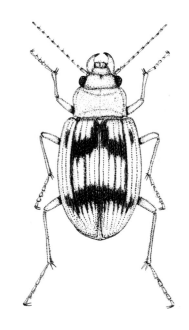

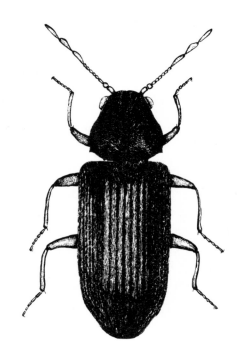

above: Great Diving Beetle, *Dytiscus marginalis*. Measures from 26-32 mm in length.

below: Inquisitive Longhorn, *Rhagium inquisitor*. Length from 12-16 mm. Found in most parts of the country where there are rotting tree stumps, principally conifers, but those of Birch and Oak in addition.

above: Common Woodworm, *Anobium punctatum*. From 2.5 to 5 mm in length. A serious pest of wood in buildings everywhere.

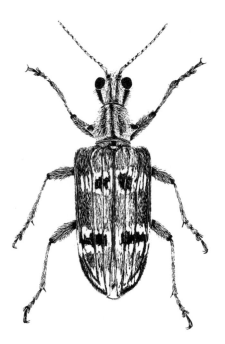

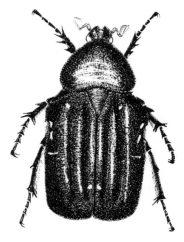

Rose Chafer, *Cetonia aurata*. From 14-20 mm in length. Occurs throughout the country.

ground colour with seven fine, black, longitudinal lines on each wing case are ornamented with two black, transverse bars of characteristic pattern. It has a curious history and distribution. Recorded on a number of occasions between 1857 and 1902 on the coasts of Wicklow and Wexford, it was not then reported until 1967 when a number were found on certain beaches in Wexford. It still occurs there today. Its habitat requirements are, firstly adequate strand-line debris; secondly, an extensive sand-dune area behind the beach and, lastly, only moderate human disturbance. These requirements appear always to be shared with the woodlouse *Armadillidium album*. In England it occurs only in North Devon and is found also in South Wales.

Great Diving Beetle
Dytiscus marginalis
The family DYTISCIDAE or water beetles, consist of a large number of species, usually small and often difficult to identify. They are all carnivorous in habit — in both the larval and adult stages. Together with the ground beetles or CARABIDAE, they constitute the sub-order *Adephaga,* which is considered to be the most primitive and least specialized of the order *Coleoptera.*

Beetles of the genus *Dytiscus,* however, give no problems of identification. *D. marginalis* is large in size, measuring from 26 to 32mm in length. The general colour is dark-olive-green with a yellow border to the pronotum and the outer margin of the wing-cases. There is a curious difference between the sexes. In the male, the first three basal segments of the first pair of legs are much enlarged, have a margin of hairs and are covered with suckers. Two of the latter are much larger than the others. The female is without these sucker-pads. On the other hand, although the male's back is smooth and shiny, that of the female is dull and the wing-cases deeply furrowed or ridged longitudinally, and these markings extend to about two-thirds of the length of the wing-cases.

The Great diving beetle, *Dytiscus marginalis,* is generally found in deep, weedy pools almost throughout Europe — in addition to North America. It emerges from mid-summer to late October, and has a long adult life. The larva is perhaps more voracious than the adult, and will kill any aquatic animal up to the size of a small fish! It is often found to be infested with the larvae of water mites, *Hydracarina.*

The Great diving beetle can often be seen resting at the water surface in an oblique position with the space between the upper surface of the abdomen and the lower surface of the wing-cases in communication with the air whilst gaseous exchange takes place. On completion of this process or when disturbed, it dives to the bottom with vigorous beats of its strong hind legs. It appears to be widely distributed in Ireland.

Common Woodworm
Anobium punctatum
Common woodworm is a serious pest of wood in buildings, both in the structural timber and in the furniture. It is known as Common furniture beetle. It does more serious harm to joists and floorboards than to furniture. It is fairly easy to recognise. It varies in length from 2.5 to 5.0mm and in colour from a light reddish-yellow through dark, chocolate-brown to pitchy-red. Some of its colour variation is, however, due to a scale-like powder rubbing off. If the beetle is viewed from the side it is seen that the head is inserted into the first thoracic segment — the prothorax — vertically downwards. This is a common characteristic of all the species of the family. ANOBIIDAE, to which it belongs. Each wing-case bears nine longitudinal rows of dark-coloured spots, which are, in reality, pits in the surface. The last three segments of the antennae are enlarged.

Soon after mating, the female seeks egg-laying sites, such as rough-end grain, and crevices or joints of wood. The eggs hatch in a few weeks and the larvae burrow straight into the wood. For about three years the larvae then tunnel mostly in the sapwood, before they turn towards the outer surface of the wood. Just before the outer surface is reached, however, the larvae construct a large chamber in which the pupal stage takes place. The pupa is at first white, but after a few weeks it becomes pigmented, and finally the pupal skin is cast and the adult, after resting for a few days, bores its way out of the

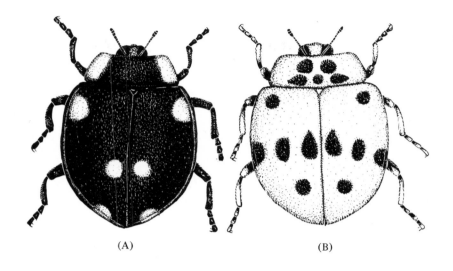

(A)   (B)

Ladybirds, COCCINELLIDAE
(a)   Two-spot Ladybird, *Adalia bipunctata*
(b)   Ten-spot Ladybird, *Adalia 10-punctata*
(c)   Fourteen-spot Ladybird, *Propylea 14-punctata*
(d)   Fourteen-spot Calvia, *Calvia 14-guttata*
All are between 3 and 5 mm in length and all are relatively
common everywhere.

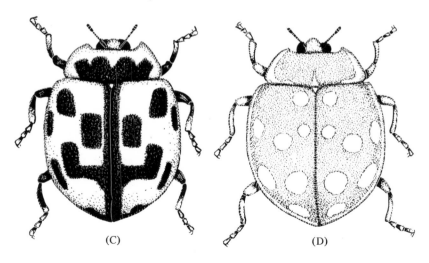

(C)   (D)

wood, finally leaving an exit or flight-hole just less than 2mm in diameter.

Woodworm prefers a temperate climate, mild and humid, and Ireland just suits it! Found throughout Europe, South Africa, New South Wales, Tasmania, New Zealand and North America, the woodworm is not found in tropical regions. The main emergence period is during the latter half of July, but in centrally heated premises it may be seen at almost any time — from spring to autumn, and occasionally in winter. The adult beetle lives for a few weeks, at most.

Ladybirds *Coccinellidae*

Ladybirds are probably the best known of all beetles. Their bright, spotted and conspicuous colouration, as well as their apparent 'tameness' makes them very popular, especially with children. Added to this is the almost universal knowledge of their beneficial character in devouring aphids, both as adults and larvae. The COCCINELLIDAE, to which ladybirds belong, is a large family — about 3,500 species in all. They are generally easy to recognize. The top surface is hemispherical whilst the undersurface is flat. The head is partly concealed beneath the prothorax and the last three segments of the antennae form a club. The legs are short and retractible beneath the body. There appear be only three torsal segments, but a minute observation shows there to be four.

The bold colouration of many ladybirds warns birds and other predators that they are distasteful and this gives them a great advantage — enabling them to walk freely at all times. However, in spite of this bright patterning, the common species still prove most difficult to identify. This is because many vary tremendously — in fact they vary so much that they can be confused with other species which, one would have thought, were distinctive enough.

Four species, however, are illustrated to give some idea of the variations. The Two-spot ladybird, *Adalia bipunctata*, is one of the most variable of insects in its colour pattern. Varying from black with red spots to red with black spots, its correct identity is a job for the trained entomologist. The Ten-spot ladybird, *Adalia*

10-punctata, is similarly variable but is *always* brown underneath — with a whitish breast and pale legs — compared with the Two-spot, which is always black underneath. The 14-spot Ladybird, *Propylea 14-punctata*, has a yellow background colour with a pattern of black spots and bars, as shown in the illustration. However, they can vary from being almost all yellow — except for the black longitudinal bar where the wing-cases meet in the middle — to practically all black — except for a narrow yellow line or marginal band around the outer wing-cases, and a central yellow spot where the wing-cases join the prothorax. The last species illustrated is the 14-spot Calvia, *Calvia 14-guttata*, which is not nearly so variable as the foregoing three species. It has an orangey-brown background with seven yellow spots on each wing-case and there are no black markings on the top surface. Of the species illustrated this is probably the rarest in Ireland.

The Rose Chafer *Cetonia aurata*

This very beautiful beetle, bright, glistening, golden-green in colour has a few small white marks across the wing cases. The latter are somewhat variable in extent — the Cork specimen illustrated appears to have fewer marks than are generally possessed. The underside is a shining, metallic copper. This species is specially attracted to flowers, roses being often chosen.

Unique amongst all beetles is its method of flight. In all other beetles the wing cases are raised at about right-angles to the abdomen to allow the wings to be released and flight to take place. In the Rose Chafer, however, the outer edge of the wing cases, near the base, are cut away and, indeed, the wings may be seen through the cut-away part. The wings can be protruded through the aperture thus formed, and flight takes place with the wing cases covering the abdomen.

The Rose Chafer is classified in the family SCARABAEIDAE, which in turn is placed in the group *Lamellicornia*. About 25,000 species of scarabaeids are known and this is one of the largest families of beetles. In addition, a number of tropical species are of outstanding size, being amongst the largest of all insects. One of the characteristic features of the *Lamellicornia* is the special form of the antennae. The last segments are flattened and can be extended to give a fan-like organ from which the name has been derived. In some years the Rose Chafer is not uncommon in the adult stage, though whether they are migrants or of local origin is debatable.

The Inquisitive Longhorn
Rhagium inquisitor
The CERAMBYCIDAE or LONGICORNIA (or just simply Longhorns), constitute the beetle family of about 20,000 named species. Members of this family are found throughout the world, wherever trees or bushes grow, and indeed wherever timber is transported or used. The number of Irish indigenous species is probably less than sixty, but many more species are found in timber (in the larval stage) which has been imported. In America there is a multiplicity of species, and, more especially in South America, some are of gigantic size and often remarkable also in form and colour. The great majority of the larvae of CERAMBYCIDAE live in and consume the woody tissue of trees and shrubs. The extent to which fungal decay occurs in the woody tissue largely determines which species of CERAMBYCIDAE will attack it.

On the whole, the CERAMBYCIDAE are rather more than medium in size, but some species are amongst the largest known insects. Their form is fairly well defined, although there is a wide variation within the family. The body shape is elongate with the wing-cases wider than the pronotum. The most characteristic feature, however, shared by all but a few species, is the extreme length of the antennae, being usually as long as or longer than the body. In one tropical species they reach a length of 22cm! The eyes are large and frequently bow-shaped and the large tubercle from which the antenna arises is located in the concavity. There are five tarsal segments but the fourth is much reduced — only four segments being usually visible. The third segment has a definite bilobed appearance.

In the Inquisitive Longhorn, the antennae are only moderate in length. It is easily identified by two black marks on each wing-case surrounded by yellowish bands. The larvae feed on the wood of conifers but they have also been collected from a number of hardwood trees, such as birch, *Betula*; and oak, *Quercus*. The author's first recollection of this species was when, as a boy wearing a white sweater on a hot July afternoon, a number of Longhorn beetles alighted on him. There have been no recent records in Ireland.

Red-spotted Weevil
Otiorrhynchus ruropunctatus
The beetle family CURCULIONIDAE, popularly known as 'weevils', is one of the largest, natural groups in the whole of the animal kingdom. It is believed that there are probably about a quarter of a million species although only a fraction of this number has been described. Members of the family are fairly easy to identify — with the head prolonged into a long snout or rostrum, at the tip of which the mouthparts are situated. The antennae are elbowed and the end segments form a club. There are apparently only four segments on each tarsus, but the true fourth one is minute and not easily distinguished. The third segment is bilobed. The tibiae are arched. Many weevils are wingless.

The larvae are legless and crescent-shaped. They never feed in an exposed situation and are always vegetarian. Most weevils are dark in colour, as in the illustrated specimen, from Co. Louth. The prothorax and wing-cases are waisted, randomly pitted. The insect is 6mm in length. The Red-spotted weevil is part of the so-called Lusitanian element of the Irish fauna, the only other part of Europe where it is found is the Pyrenees.

Giant Crane Fly *Tipula maxima*
The large family TIPULIDAE represents the most primitive and least specialized of the true flies — the order *Diptera*. The adults are known as crane flies, or more popularly as 'Daddy-long-legs', on account of their extremely long legs. The larvae, however, are better known to farmers, gardeners and groundsmen as serious pests of grasslands, and are known as leather-jackets. These are grey and legless with tough, rubbery skin and with biting mouth-parts. They live in the soil and feed on the lower stems of a

wide variety of plants, including cereals and grasses. They may also feed on the surface at night in suitable weather conditions.

The adult Crane fly has a long, slender body with prominent genitalia at its apex and a single pair of narrow, delicate wings and long, fragile legs. The highly modified, indeed extremely reduced, second pair of wings consist only of club-like organs known as 'halteres' and are quite conspicuous. The Giant Crane fly is the largest member of the Irish flies, the length of its wing-span being 65mm. The body length is approximately 32mm. The larvae can be said to be amphibious, and may be found in the mud and vegetation around ponds, where it can remain under-water or in water-saturated conditions by plugging its air-breathing spiracles with gill-like tentacles capable of absorbing oxygen from the water. When breathing air, however, the larva is seen with its tentacles in a rosette-like structure around the spiracles.

The White-banded Drone Fly
Volucella pellucens

This is another large fly which should present no problems in identification. It is about 15mm in length with the largest abdominal segment near the thorax white and conspicuous, whilst the thorax and the rest of the abdomen is shining black with a number of black bristles. The wings have a broad, smokey-brownish-black band across the centre and the tips are smokey-brown also. The bases of the wings are suffused with bright orange. The males are often observed hovering in almost stationary flight (somewhat over two metres from the ground) in woodland glades, whereas the females are most often to be

White-banded Drone Fly, *Volucella pellucens.* About 15 mm in length. Probably generally distributed.

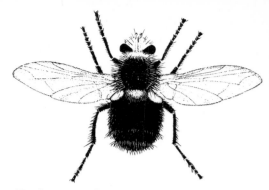

The Great Black Parasite Fly, *Larvaevora (Echinomya) grossa.* About 32 mm across the outstretched wings. Specimen illustrated from Co. Cork. Its distribution and abundance not known in Ireland.

found sucking at flowers of bramble and dog rose for nectar.

The immature stages are spent in the nest of the Common wasp, *Vespula vulgaris,* where the eggs are laid and suffer no harm from the wasps. The larvae act as scavengers, feeding on the dead bodies of wasps and larvae, as well as on excreta. Wasp larvae do not normally excrete until shortly before pupation, but the Drone fly larvae are able to induce them to do so and thus provide a liquid meal when it is required. The SYRPHIDAE, the family of flies to which the White-banded drone fly belongs, are known as hover flies, drone flies or flower flies. They are characterized by their hovering flight, their bright coloration (often resembling wasps or bees), and their great attraction to flowers. The larvae of some species feed on aphids, but many types of life history are to be found in the group. 165 species of Syrphid flies occur in Ireland.

The Great Black Parasite Fly
Larvaevora (Echinomya) grossa

This very large, black fly can be easily identified. It is about 32mm across the outstretched wings, and the upper surfaces of the thorax and abdomen are black and covered with black bristles. These give the fly a characteristic 'black bristly' appearance. Furthermore, the head (with the exception of the dark-coloured eyes) is yellow. It is usually observed sitting on the flowers of UMBELLIFERAE and COMPOSITAE. It is common and widely distributed in southern Ireland. The specimen illustrated was collected in west Cork. In Britain it is confined to the

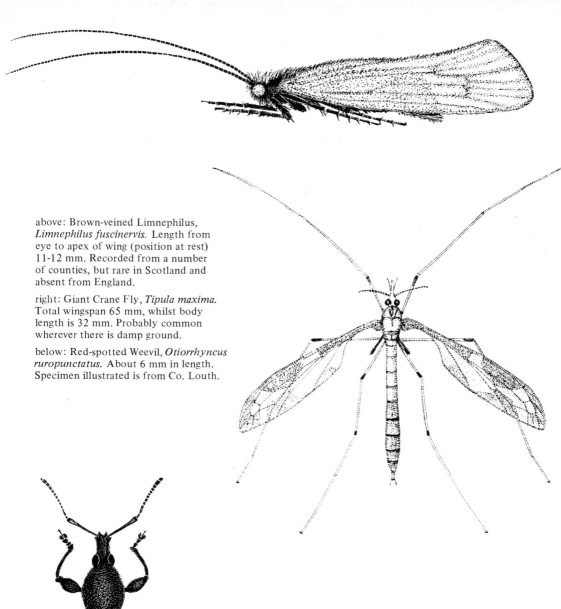

above: Brown-veined Limnephilus, *Limnephilus fuscinervis.* Length from eye to apex of wing (position at rest) 11-12 mm. Recorded from a number of counties, but rare in Scotland and absent from England.

right: Giant Crane Fly, *Tipula maxima.* Total wingspan 65 mm, whilst body length is 32 mm. Probably common wherever there is damp ground.

below: Red-spotted Weevil, *Otiorrhyncus ruropunctatus.* About 6 mm in length. Specimen illustrated is from Co. Louth.

counties of Devon, Cornwall, Hants and Dorset. The larvae are known to parasitize the larvae of large moths such as the Fox moth, *Macrothylacia rubi*.

The family to which this species belongs, the LARVAEVORIDAE, was formerly called the TACHINIDAE. About 250 species occur in the British Isles and all are parasitic on the larval stages of insects or on centipedes, spiders and woodlice. Predominantly, however, they are parasites of the caterpillars of moths and butterflies. The barrel-shaped larvae live inside the body of their host, feeding on non-vital tissues until shortly before being fully grown, when all the tissues are consumed and the host larva is left as a lifeless skin.

Brown-veined Limnephilus
Limnephilus fuscinervis
Caddis flies, *Trichoptera*, are moth-like flies closely related to the *Lepidoptera* — moths and butterflies. Instead of wings being covered in scales, as in the latter group, the wings are more or less hairy. There is no long-coiled proboscis and the bright, conspicuous colouration of the wings does not occur, indeed their wings are of a somewhat sombre hue. However, the greatest difference between these insect orders lies in their biology. Almost without exception the immature stages of moths and butterflies are terrestrial. Those of the Caddis flies, on the other hand, are aquatic. The Caddis fly lays her eggs in or very near the water, and the young larvae enter (or are already in) the water, and during the whole of this and the subsequent pupal stage, obtain oxygen for respiration direct from water through the medium of gills, of which there is a variety of types. There is great variation also in the larval form and habit.

Perhaps the best known Caddis flies belong to the large family, LIMNEPHILIDAE. In these the larval abdomen is white and unprotected, but a case is constructed in which the larval stage is spent. The material used for the case as well as the pattern is often characteristic of the species, and the habitat — whether slow-flowing rivers, mountain streams or moorland trickles and species of vegetation — is also of great interest. The pupal stage of Caddis is agile. It is

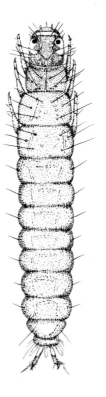

Wet Rock Tinodes, *Tinodes assimilis*. Length at rest 7 mm. Recorded from Tramore, Co. Waterford in 1952 but now absent. It must occur elsewhere in Ireland!

furnished with functional mandibles by means of which it cuts its way out of the cocoon and by using its fringed legs, swims to the surface and crawls onto emergent vegetation or stones in order to metamorphose into the adult fly.

The first list of Irish Caddisflies was made in 1910, when 114 species were recorded, but it was not until 1972 that serious work commenced to get something like a complete view of this aspect of the fauna. However, enough collecting and identification has been carried out to realize that there are unusual anomalies. Several species which would have been expected to be present have, so far, not turned up, whereas other species appear abundant, and yet are rare or absent in Britain. The species which has been illustrated is the Brown-veined limnephilus. It measures from 11 to 12mm in length and is of a straw colour with no markings, but the veins are conspicuous being dark-brown when seen with a x10 lens. It is absent from England and of rare occurrence in Scotland. In Ireland, however, there are records of it from Co. Mayo and Co

Monaghan in 1909 and 1916, but it has been recently collected (1975) in Cavan, Clare and Offaly.

The Wet Rock Tinodes
Tinodes assimilis

By no means all the larvae of Caddis flies construct cases in which to live and which they drag about with them. A considerable group inhabiting fast-flowing streams and torrents — notably the family HYDROPSYCHIDAE — spin webs. Here they lie in wait for animal or vegetable organisms of small size to become entangled, and then they emerge from a bag-like extremity of the web and consume them. Some species are free-living and appear to wander around mossy stones in quick-flowing water, with apparent immunity from predators, to see what they can pick up — these belong to the family RHYACOPHILIDAE.

The Wet Rock tinodes, however, belong to a family PSYCHOMYIIDAE whose species make long, tunnel-like structures of a secretion somewhat akin to silk. This adheres along the whole of its length to the substrate — stones, sunken tree branches etc. The larvae never leave this rather elongate retreat in which they conceal themselves very efficiently, but make a cocoon near the mouth of the tunnel from which the adult Caddis fly eventually emerges.

As far as is known, the Wet Rock tinodes is the only Caddis species which has an environment consisting only of the wet film of water running down rock surfaces. The animal species inhabiting this very specialized environment or ecological niche is known as the 'fauna hygropetrica' and includes one dipterous fly, *Thaumaulia testacea*, whose larvae moves through the film by alternately flexing its body. The long, fixed galleries are about 5 to 6mm in length and about 3 to 4mm in width. They are composed of secreted silk to which detritus of various sorts is attached, and occupy irregularities in rock surfaces over which a thin film of water is running. The adult is small, being only about 7mm in length when at rest. The writer was able to record this as an Irish species when, in 1952, it was found on about a square metre of wet rock surrounded by pavements and brick

walls in the centre of Tramore, Co. Waterford. In 1977, however, a survey of the locality showed that the wet rock had been bricked over — presumably to tidy it up — but surely this species must occur elsewhere in Ireland!

Giant Wood Wasp *Urocerus gigas*

The order *Hymenoptera* consists of the ants, bees and wasps. It is without a doubt the most highly specialised group of insects. A most important set of features serve to identify them from insects which may superficially resemble them. There are two pairs of membraneous wings — the first pair being larger than the second and yet are interlocked together by means of a row of hooks. The mouthparts, although primarily adapted for biting, may also be used for lapping or sucking. Except for the less developed wood wasps and saw flies, the abdomen is constricted at the first segment which is fused with the thorax. A more or less conspicuous ovipositor is present and this is modified for sawing, piercing or stinging. A complete metamorphosis takes place — the larval stage being legless except for the more primitive saw flies, whose larvae may sometimes be mistaken for those of the *Lepidoptera*.

Perhaps the most important feature, however,

Giant Wood Wasp, *Urocerus gigas*. Females may reach 50 mm across outstretched wings but males are known of only 18 mm. Generally distributed where there are old, coniferous trees but often seen in new buildings where emergence has taken place from imported softwood timber.

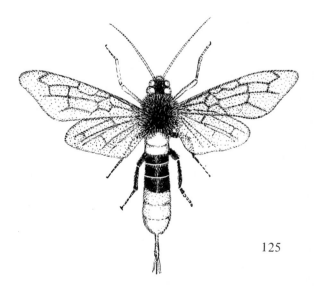

of the group is that of social behaviour, shown particularly in the ants and the honey bee — although the development of the various degrees of this social behaviour can be traced in the group generally. Two sub-orders are easily recognizable. In the *Symphyta* there is no marked constriction in the abdomen, and the larvae have three pairs of thoracic legs, and often six or more pairs of abdominal limbs. They are plant feeders. In the *Apocrita,* on the other hand, the insects are 'wasp-waisted' and the larvae legless, and they are mostly carnivorous.

An example of the order is the Giant wood wasp — a member of the *Symphyta.* The family SIRICIDAE contains a number of species important as pests of timber. It is doubtful whether any species are indigenous to Ireland as the only native conifer, *Pinus sylvestris* died out in the Middle Ages, but the larvae travel about the world in consignments of commercial softwoods. The Giant wood wasp may reach a length of 50mm but this is very variable, and males as small as 18mm are known. The pronotum is black and the central abdominal segments are black also. The remainder of the abdomen is yellow, and the ovipositor and its accessories are elongate. This gives to the insect a remarkable similarity to a hornet, so that when it emerges from new wood in a building, it often causes alarm, although unnecessarily as it is entirely harmless to humans.

This insect also has a fascinating life history. The female examines the bark of a tree (usually Scots pine, silver fir, larch or spruce) with the antennae, then she makes a few trial borings before settling down to oviposition. This is accomplished by the ovipositor sheathes being pressed against the bark, whilst the flexible ovipositor is forced into the bark and moved up and downwards in a sawing movement. At the bottom of the resulting shaft an egg is laid. The female then withdraws the ovipositor a little and lays another egg. The eggs and tunnel wall are covered with a glandular secretion which is white, glistening and glue-like, and is apparently a lubricant to facilitate the egg passage, but may also contain the reproductive stages of fungi which will eventually decay the wood in which the larvae will feed. Special organs are developed in the larvae which ensure the survival of the fungus. Associated with the Giant wood wasp are a number of insects parasitic on it. Altogether the biology of this insect makes an extraordinary chapter in insect natural history.

11 Millipedes and Centipedes

Millipedes *Diplopoda*

The millipedes constitute a separate class of the *Arthropoda* known as the *Myriapoda*. Their most important characteristic is two pairs of legs on most segments. There is a distinctive head which bears a pair of antennae which are short and composed of eight segments; a pair of mandibles and a plate beneath them which is made up of a pair of fused mouthparts — the maxillae. Eyes are present in some families. Respiration takes place by means of a network of tubules (tracheae), which lead from openings (spiracles) at the base of the legs. The upper parts of the segments are much more developed than the lower parts. In some of the hinder segments legs are absent, and the shape on the last segment is often important in identification.

The apertures of the reproductive organs are on the underside of the body, near the head and this contrasts with the centipedes, where these organs are found on the last segment of the body (as in insects). Although sometimes recorded as feeding on various dead soil animals, they are predominantly vegetarian and occasionally pests of crop plants.

The species illustrated, *Brachydesmus superus,* is a member of the family POLYDESMIDAE. The latter consists of millipedes with flat segments which are extruded laterally into 'keels'. The top surface of the segments bears an intricate design and eyes are absent. There are nineteen body segments in *B.superus,* and they are light-brown or of a creamish colour. In length it is from 8 to 10mm, and in width 0.8 to 1.00mm. It is an inhabitant of farmland soil, but is also known from caves.

Centipedes *Chilopoda*

This is another and quite separate class of the *Arthropoda*. Compared with the millipedes, centipedes are predaceous. The head is distinct and the body, which is long and narrow, is divided into an extremely variable number of segments. Indeed, whilst some species have fifteen, others may have well over one hundred pairs of legs. Except for the first, all of the segments are provided with a pair of legs. The head bears a pair of many-segmented antennae; a pair of toothed jaws and a pair of leaf-like maxillae. The first body segment bears a pair of poison claws which are used for capturing and killing prey.

Eyes are present in many species, and consist sometimes of a group of simple eyes (ocelli) or are compound — like those of many insects. The number of eyes and their arrangement is often useful in identification. The flattened body segments consist of upper and lower horny

Millipede, *Brachydesmus superus*.
Length 8-10 mm. Probably
common in farmland.

plates, united by a membranous skin with which
the legs articulate and through which respiration
takes place via openings of the tracheal system
(spiracles), as in millipedes. Centipedes lack the
waterproofing layer on the skin as in insects, so
that they are always vulnerable to drying out in
sunlight. For this reason they always hide
amongst damp litter, in damp soil and under
stones or similar situations — always shunning
the light.

Brachygeophilus truncorum

This long and very slender centipede is contained
within the family GEOPHILIDAE. Varying in
length from 12 to 14mm and up to 0.6mm in
width, it has from 37 to 39 body segments in
the male, whilst the female may possess one or
two more. The head and first body segment are
a little darker in colour than the pale yellow to
light-brown general body colour. This species is
generally found in woodland areas but also
beneath the bark of dead and decaying trees in
other situations. It is generally abundant where it
occurs. In Ireland, it was previously recorded
from Antrim, Down, Louth, Monaghan and

Tyrone, but the specimen illustrated was
collected from a garden in Sandymount, Dublin.

Lithobius forficatus

This species is a common European centipede
and perhaps one of the more-widely known
because of its association with the immediate
surroundings of buildings and suburban gardens.
Indeed, its propensity for nocturnal wandering
often finds it under a doormat in the morning.
It is not entirely dependant on man, however,
as it commonly occurs far away from human
habitation in woods, grassland areas and on
moors, up to an altitude of about 500m from
just above high-tide mark. It is most usually
found, however, under stones and in decaying
timber. It is a fairly large species and may attain
a length of 32mm and a width of 3.8mm. In
colour it is dark-brown to chestnut. The head is
a little wider than it is long, and the antennae of
35 to 43 segments occupies a length of about
one-third of its body length. There are from 20
to 30 'eyes' (ocelli) on each side of the head.

Its known distribution in Ireland rather
reflects the distribution of people able and
willing to record it; it is likely to be found in all
parts of the country. Looking at distribution
records, as far as they exist, this could be said of
Ireland and Britain as a whole. Abroad, it is
found in Northern Europe as well as in the
Mediterranean region and North Africa. As an
introduction, it has become established in North
and South America, and also in the Atlantic
Island of St. Helena. Another species in this
genus, *L. variegatus,* has wide distribution in
Ireland. It has a distinctive colour, being pale-
brown variegated with violet. The yellow legs
are banded with pale brown. It is not known
outside Ireland and Britain (including
interestingly, as far south as the Channel Islands).

Two further species often found in Irish
gardens are: *Lithobius microps* and *Cryptops
hortensis.* The former is a small brown species
that curls up when disturbed whilst the latter is
brownish, has twenty-one pairs of legs and is
very fast-moving.

My collection of spiders, centipedes and
millipedes was kindly identified for me by
Professor John L. Cloudsley-Thompson.

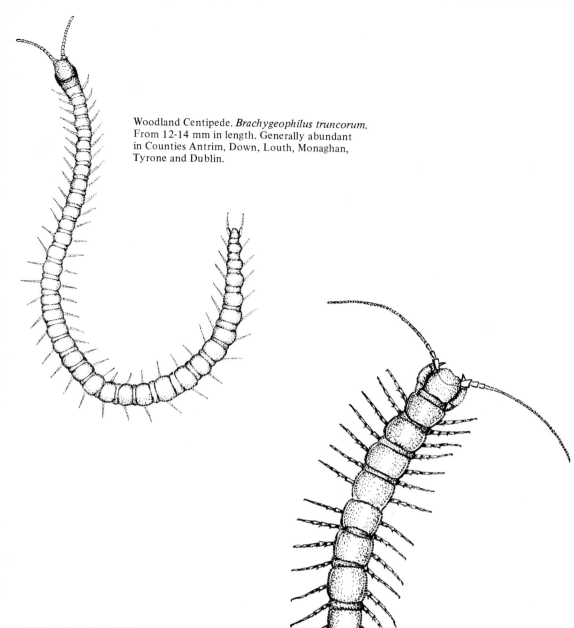

Woodland Centipede. *Brachygeophilus truncorum.*
From 12-14 mm in length. Generally abundant
in Counties Antrim, Down, Louth, Monaghan,
Tyrone and Dublin.

Garden Centipede, *Lithobius forficatus.* Up to
32 mm in length. Predominantly recorded from
coastal counties, except those of the South-east.

12 Crustacea

Barnacles, Shrimps, Prawns, Crabs, Lobsters, Crayfish, Woodlice

Due to the wide range of form in this group, difficulties arise in trying to give a simple description. *Crustacea,* however, are aquatic arthropods (but woodlice are terrestrial), which breathe either by gills or through the general body surface. There are two pairs of antennae arising from the head. The body is segmented, although the segmentation of the head and thoracic region is often indistinct because of large flaps which extend backwards, forming the 'carapace'. Each segment bears a pair of appendages which are modified in various ways in order to perform a number of different functions: antennae; mouthparts; seizing legs; walking legs; swimmerets; to the segments forming a tail fan (telson). There are many variations from the basic plan, however, and in parasitic forms great reduction in complexity is often shown.

Classification of the *Crustacea*
The *Crustacea* is a complex and diverse group made up of over thirty major divisions. Unfortunately, little space can be devoted to classification in the following brief accounts of a few species, and none to the great variation in immature forms. The student however will find much of interest in further reading on these subjects.

Goose Barnacle *Lepas anatifera*
Barnacles are grouped together as the crustacean sub-class *Cirripedia.* They have a most un-crustacean-like appearance — indeed, they were thought to be molluscs, until 1829 when the immature stages were found. These resembled the larvae of crabs, and so their true relationships were established. The Acorn barnacle, *Balanus balanoides,* occurs throughout the rocky shores of Europe in the greatest abundance. It has been calculated that a 1.6km (1 mile) length

Goose Barnacle, *Lepas anatifera.* About 30 mm in length excluding appendages. Occurs from time to time attached to floating timber.

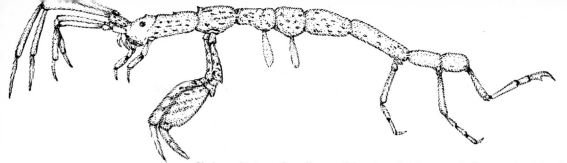

Skeleton Shrimp, *Caprella aequilibra*. Length 15 mm excluding anterior 'claws'. Clinging to seaweeds probably widespread and plentiful.

of rocky shore may contain as many as 2,000 million Acorn barnacles!

The soft body is contained within a shell which is cemented to the rock. The aperture of the shell can be closed by four valves attached to a flexible membrane. When the aperture is opened, six pairs of whip-like cirri emerge and make grabbing movements to catch small organisms for food. The water currents are also set up to aid respiration. The five posterior pairs of cirri correspond with the five pairs of walking legs of the more advanced crustacean groups.

below left: *Conilera cylindracea*. About 20 mm in length. Probably widespread around the coast in seaweed beds.

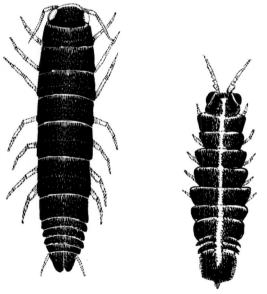

above right: *Idotea baltica*. About 15 mm in length. Probably common everywhere from low tide mark into shallow water.

The Goose barnacle belongs to the 'stalked' barnacles — the *Pedunculata*, where the soft body of the animal is enclosed by a number of calcareous plates borne on a stalk. From time to time they are found floating on timber, so that the beachcomber naturalist should examine all such pieces of wood whenever possible by turning the wood over carefully, in case there is a colony on the other side. The author once walked along a thrown up plank covered with long strands of seawood of a greenish hue and, on turning it over, found to his dismay that a Goose barnacle colony was crushed completely on the other side.

A word must be spared for the mention of the old Goose barnacle myth which comes down to us from the Middle Ages. It was not until 1891 that the nesting habits of the Arctic-breeding 'Barnacle Geese' were made known. Before that time there were stories of the mediaeval writers whose explanations were legion, but were all agreed that the 'Barnacle Geese' were produced from the barnacles found on floating timber. There were strange similarities, but all now exploded in legend.

Skeleton Shrimp *Caprella aequilibra*

The Skeleton or Ghost shrimp is classified in the *Amphipoda*, one of the most important groups in the *Crustacea*. There are thought to be well over 3,600 species known from the seas of the world, and they are present in immense numbers. Although there is a predominance of marine species, nevertheless they extend through brackish to fresh water. This species belongs to the sub-order *Caprellidea*, although the *Gammaridea* are more representative of the group as a whole.

131

The typical amphipod has a body showing distinctly the segmentation and division into a head, thorax and abdomen. This is on account of the carapace, present in the decapods, being absent. The first thoracic segment, however, is fused to the head, leaving seven free segments in the thorax. There are six distinct abdominal segments and there is a free terminal segment. Usually the body is flattened from side to side, arched, with a deep groove between the legs. The thoracic legs do not support the body so that amphipods do not walk 'upright' but scramble about sideways. Anyone who has turned over some damp seaweed on the shore must have observed the *Gammarus* shrimps jump away sideways to seek cover elsewhere.

An interesting, but harmful amphipod species is *Chelura terebrans,* which is associated with the isopod crustacean *Limnoria lignorum* (the gribble) which tunnels into unprotected wooden harbour works and jetties, as well as into wooden-hulled boats.

Conilera Cylindracea
Conilera cylindracea is given here as an example of the *Isopoda* group of the *Crustacea.* It is an external parasite of fish, about 2cm in length and is often found amongst seaweed at low tide. Isopods are even more numerous in world species than are the amphipods — about 4,000 species being already known, and many still await scientific description. There are many terrestrial forms, such as the well-known woodlice. Some isopods inhabit fresh water but the great majority are marine.

In contrast to the amphipods, isopods are flattened from the top downwards, giving an oval outline which may be exaggerated outwards by extensions of the segmental plates. Like amphipods, however, the body is clearly divided into head, thorax and abdomen, the head furnished with stalkless, well-developed eyes. The thorax is eight-segmented, although the first (and sometimes the second segment also) may be fused to the head. The abdomen is six-segmented, and usually the segments continue on from those of the thorax without any clear demarcation. The last abdominal segment is fused with a fan-like

tail. There are five pairs of head appendages — the antennae often being very long. Generally isopods are scavengers and omnivores, but some are herbivores, eating seaweed.

Idotea baltica scrapes and bites off pieces of seaweed with its specialized mandibles. There are seven pairs of thoracic 'walking' legs, all similar in shape — *Isopoda* means equal-footed. The first five pairs of abdominal appendages are two-branched and leaf-like, serving a respiratory function.

Dublin Bay Prawn *Nephrops norvegicus*
Rather like a small lobster, the Dublin Bay prawn — also known as the Norway lobster — is found throughout the shallow and deep water (about 250m) around Ireland. It is commonly distributed elsewhere from the south coast of Ireland, the Norwegian coast, Scotland, western France, Spain, Portugal and the western Mediterranean. The body length, excluding claws, is up to about 24cm for the male, and 15cm for the female and they are rosy-grey in colour. They subsist on small invertebrates which are hunted at night — the day usually being spent buried in mud or sand. They are very slow-growing.

Before the early 1950s, they were usually discarded as human food, but since then the demand for langoustine in France and scampi in Italy (and elsewhere!) has increased tremendously, and now some countries have minimum catching sizes of 11-16cm. It is included in the Official United Nations (FAO) Catalogue in which it is numbered 512.

The annual catch for Ireland is about 2,000 tons and it is now a 'gourmet's' dish. According to Elizabeth David, the delicate flavour is perhaps best savoured if the tails are cooked, still in the shell, in salted water for ten minutes, and then the prawn served hot with melted butter.

Burrowing Prawn *Upogebia deltaura*
This decapod is rare in Irish waters, although what few specimens have come to light have been well distributed around the coast. It forms a burrow in sand in littoral or shallow water, but

emerges to feed. Some specimens have been retrieved from fishes stomachs. It is lobster-like in general form, but the first two abdominal segments are small, giving the body a 'waisted' appearance.

Hermit Crab *Pagurus bernhardus*

This Hermit crab is common all around the Irish coast in shallow to deep water. It is widely distributed throughout the European seas. The Hermit crabs, of which nine species are found in Irish seas, are classified in the decapod family, PAGURIDAE, and although the head, claws and first two pairs of walking legs have a lobster-like appearance, the carapace and abdomen are very soft and much atrophied. The abdomen twists to the right, and occupies the empty shell of a sea-snail — the larger specimens of the present species are generally found in the shell of the whelk, *Buccinium.*

There are no appendages on the right side of the abdomen except for a part of the tail-fan, and this sticks out, preventing the crab from being pulled from the shell. This tail-fan is covered with spines so arranged that the harder it is pulled the more firmly the crab clings! The last two pairs of walking legs are reduced in size and serve to help carry the shell when moving. When danger is sensed, the crab is able to retire well into the shell, and only the tips of the pincers are visible and these are usually well-armoured. Most of its food is obtained by scraping the algae and animal life off rocks, but it is capable of breaking barnacles and tube-worms open, devouring the occupant. Hermit crabs are well-known to live in close association with other animals in which some advantages are shared (however unfairly!) The Sea anemone, *Calliactis parasitica,* is frequently found attached to the crab's whelk shell, and the upper whorls of the shell are often inhabited by a polychaete worm, *Nereis fucata.*

Another common Irish Hermit crab occurring in shallow water is *Pagurus prideauxi.* It is generally associated with the Sea anemone, *Adamsia palliata.* This crab does not change its shell as it grows larger (as do other species) but the anemone grows forward and covers the Hermit's

soft abdomen, thus protecting it! At the same time the anemone's tentacles are situated close to the crab's mouth, and thus the anemone can take full advantage of the 'crumbs from the table'!

Toothed Crab *Atelecyclus septemdentatus*

The *Brachyura,* the sub-order of the *Decapoda,* to which this species belongs, contains the crab-like forms. The abdomen is greatly reduced, much shorter than the cephalothorax, and is permanently flexed beneath it. The second pair of antennae (antennules) and the eyes are capable of being retracted into cavities. The Edible crab, *Cancer pagurus,* is placed in this group. *Atelecyclus septemdentatus* occurs in small numbers all round the coast in littoral to deep water. In general it is crab-like but the carapace is strongly toothed along the very convexed front margin. The claws are large and strong. The specimen illustrated was 3cm across the carapace. It is closely related to *Corystes cassivelaunus.*

Masked Crab *Corystes cassivelaunus*

This species is a common Irish one found around the low-tide mark and in shallow water where there is sand. It is not, however, usually observed — except by the expert, on account of its habit of burrowing, for which it is specially adapted. If placed on sand, it quickly digs into it with its sharply-pointed 'walking' legs, and leaves only the tips of its long antennae projecting above the surface. Each of the antennae bear two rows of stiff hairs directed inwards, and when the antennae are brought together the hairs interlock to form a long tube. Water is drawn down this tube by a pumping action caused by the beating of a pair of mouth appendages. The water is stored in a large branchial chamber and is replenished, as required, by a reversal of the action. The crab lies buried by day, but emerges at night to feed. The claws are twice as long as the 'walking' legs and the carapace, which bears four sharp spines placed on each side of the front margin, is about 4cm across.

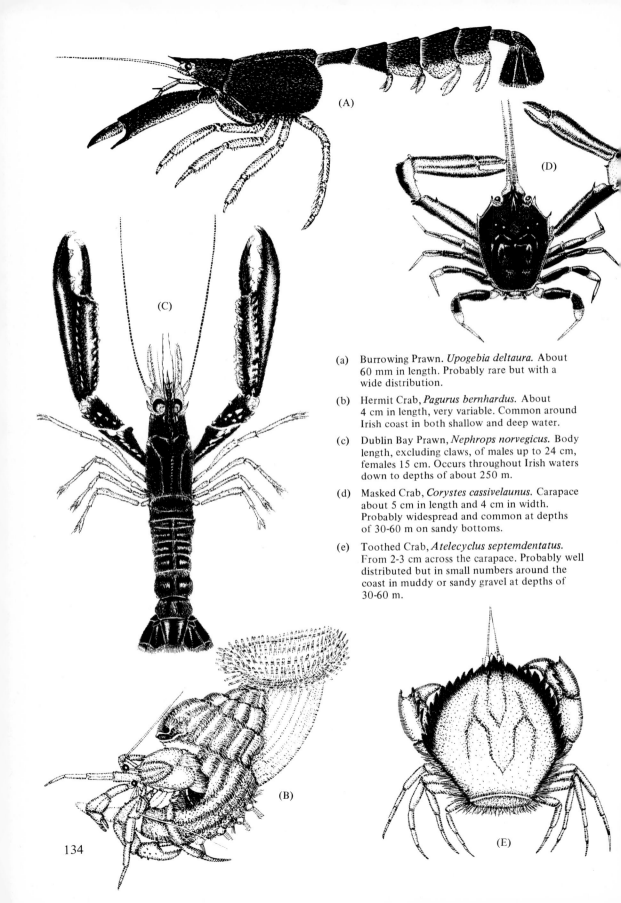

(A)

(D)

(C)

(a) Burrowing Prawn. *Upogebia deltaura.* About 60 mm in length. Probably rare but with a wide distribution.

(b) Hermit Crab, *Pagurus bernhardus.* About 4 cm in length, very variable. Common around Irish coast in both shallow and deep water.

(c) Dublin Bay Prawn, *Nephrops norvegicus.* Body length, excluding claws, of males up to 24 cm, females 15 cm. Occurs throughout Irish waters down to depths of about 250 m.

(d) Masked Crab, *Corystes cassivelaunus.* Carapace about 5 cm in length and 4 cm in width. Probably widespread and common at depths of 30-60 m on sandy bottoms.

(e) Toothed Crab, *Atelecyclus septemdentatus.* From 2-3 cm across the carapace. Probably well distributed but in small numbers around the coast in muddy or sandy gravel at depths of 30-60 m.

(B)

(E)

13 Worms

Earthworms, Polychaeta, Leeches

Worms *Annulata*

The word Vermes, from which the word 'worm' is derived was originally applied to any small, slender, legless creature. In addition to the various distinct groups of animals to which we give the word 'worm' today, this word was also used for snakes, caterpillars and, indeed, any long-bodied insect. Zoologically it is a very imprecise word, and is now only used for common names with a suitable prefix, such as 'earthworm' or used to denote a particular form. Amongst the invertebrate animals there are a number of major groups with a worm-like shape, but in this chapter examples of only the 'Ringed' worms, *Annulata,* and the leeches, *Hirudinea* will be given.

Perhaps the most important class of *Annulata* is that containing the earthworms and the marine *Annelida.* An important biological feature is shown by these animals — known as the metameric segmentation. The body consists of a series of more or less similar segments known as metameres. Externally the body is elongate and divided into a number of rings. Each ring or annulus may bear a muscular process which serves as a limb and, in addition, may bear strong bristles known as 'chaetae' which give the name of *Chaetopoda,* meaning 'bristle-footed' to this group. Corresponding to the external rings is a serial repetition of the internal organs and this is especially shown by the musculature and blood systems, and, to a lesser degree, by the renal and genital systems.

The Common earthworm, *Lumbricus,* is an example of this group, although to some extent it is not typical. It is specially adapted for burrowing. As in all *Chaetopoda,* the cuticle is thin and, unlike insects, there is no mechanism to restrict water loss (or drying). The earthworm, however, deals with this situation by secreting copious mucilage and only appears above the ground during the night or wet weather. An adaptation to burrowing is the reduction in sense organs and other appendages at the front of the body which might otherwise become damaged or hinder progress through the soil. The chaetae of the earthworm are very small but are, nevertheless, present. Any bird trying to pull an earthworm from its burrow soon finds this out.

Earthworms are exceedingly numerous and play a significant part in breaking up and aerating the soil. They feed on organic detritus in the soil as well as on dead leaves which are dragged into the burrow. The wormcasts on the surface of the soil give an indication of the presence and abundance of earthworms. They are hermaphrodite.

Sea Mouse *Aphrodite aculeata*

This marine chaetopod worm is to be found just beneath the surface of sand, with the tip of the abdomen bent upwards so that fresh sea-water is drawn downwards into the 'felt' of fine bristles in order that respiration can take place. The Sea mouse may attain a length of 20cm but about half this size is more usual. It is generally found at depths between 40 to 80m, from which it is dredged during fishing operations. The top surface is covered with bristles of several lengths and these give it an irridiscent effect. The under-surface is bare and shows the segmentation. There is a series of disc-like scales which overlap underneath the bristles on the upper surface. It feeds on other worms and other animals which are encountered during its burrowing operations.

Terebella lapidaria

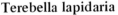

This polychaet marine worm is a common denizen of cracks and slits in rocks, just below low-water mark. It is about 25mm in length. The first ten segments bear limb-like bunches of chaetae, and, when feeding, a number of fine filaments can be extruded for entangling minute organisms. The hinder end of the body is enlarged.

Spirorbis borealis

The small, white calcareous tubes of this poly-chaet are a common sight to all who walk along our shores. The tube, which the worm inhabits, is coiled into a flat spiral up to about 4mm in diameter. It adheres strongly to stones and shells, and especially to the fronds of seaweed. The tentacles are branched and the entrance to the tube can be closed with a shield-like operculum when above water. The operculum also doubles as a brood pouch for the developing embryos. Eggs are usually to be seen through the thin cuticle near the end of the abdomen.

Leeches *Hirudinea*

There are about 500 species in the class *Hirudinea* — usually called leeches. Most are inhabitants of freshwater or are terrestrial,

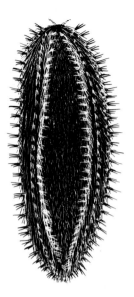

above: Sea Mouse, *Aphrodite aculeata*. Length usually about 10 cm but may reach 20 cm. Generally at depths between 40-80 m on sandy bottoms. Probably common

below: Sea Mouse from below.

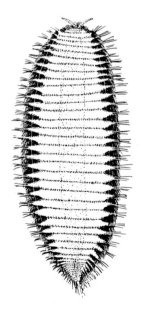

especially in tropical countries. A few, however, are marine. About three-quarters of them suck blood, and the remainder are predaceous and swallow insects and other small animals whole.

The general anatomy of leeches is much like that of the Common earthworm, but the leech is highly specialized for its mode of existence. The shape of the body is long, tapering at the front, and flattened from top to bottom. There are no conspicuous external organs and both ends are modified as suckers. The body consists of 34 segments, but of these the first five and the last seven form the suckers, and each segment is subdivided into either three or five annuli, with the anterior sucker surrounding the mouth and the posterior sucker turned downwards. Leeches do not possess chaetae.

Locomotion in leeches is brought about by the use of suckers (apart from a limited amount of swimming by 'looping'). When the posterior sucker is attached the body is lengthened. It then attaches itself by the anterior sucker, and the posterior one is then released. The body then contracts and the posterior sucker is drawn behind the anterior, which then attaches, and so on. . . In Ireland there are about twelve freshwater species and a few are exclusively marine. Amongst the former is the Medicinal leech, *Hirudo medicinalis,* which was at one time abundant but now may be considered rare. In farm ponds used by horses and cattle it fastens itself on the host and selects an area of thin skin such as the nostrils and then tightens its grip with the sucker and then proceeds to slit the skin by making rapid slicing movements with its jaws. An anaesthetizing substance is injected whilst this takes place, so that the host is unaware of it. The blood is then sucked into the pharynx, and meanwhile another substance called 'Hirudo', an anticoagulant, is secreted, thus preventing the blood from clotting.

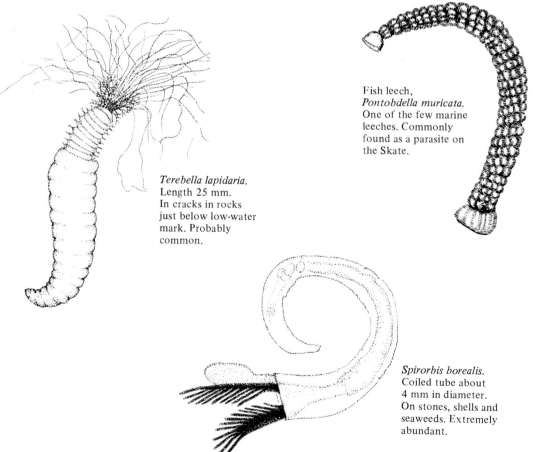

Fish leech,
Pontobdella muricata.
One of the few marine leeches. Commonly found as a parasite on the Skate.

Terebella lapidaria.
Length 25 mm.
In cracks in rocks just below low-water mark. Probably common.

Spirorbis borealis.
Coiled tube about 4 mm in diameter.
On stones, shells and seaweeds. Extremely abundant.

14 Echinoderms

Starfishes, Sea Urchins and their allies

No other group within the whole of the animal kingdom exhibits such a wide variation in external form as is found amongst the starfishes, sea urchins, heart urchins, sea lilies, sea cucumbers and their relations known collectively as echinoderms. The latter literally means 'spiny-skinned' and refers to one characteristic shared by almost all members – the presence of an *internal* skeleton of spines or plates, often fitting together in a highly complex, reticulate pattern but *within* the body cell tissues.

There is an unusual five-fold symmetry shown by the body and there is no exception within the group to this phenomenon. Perhaps most important of all, however, is the presence of a pattern of tentacle-like 'tube-feet' which operates by a special vascular system derived from extensions of the body cavity. One component of the system is easily recognized and is known as the madreporite, and consists of a small, round plate pierced by a number of pores. The tube-feet, which can be everted or withdrawn at will, are very efficient, but at the same time are most delicate and, indeed, so finely balanced that they can only operate in sea water. There are no echinoderms found in fresh water or in brackish water but, on the other hand, they are of wide distribution – occurring from tropical to polar seas, and from littoral regions to great ocean depths. The group is of ancient lineage but in Paleozoic times the sea lilies, which are stalked and fixed, were the dominant forms.

Cushion Starfish *Porania pulvillus*
The general appearance of the Cushion starfish is, in fact, like that of a five-pointed cushion, the upper surface being very convex. The five rays are triangular and very short. It is fairly large; the specimen illustrated was 15cm in greatest diameter and yellowish-orange in colour, with the rays and the centre of disc of bright red colour. It is usually found on a sandy bottom, and is of interest in that the undersurface is clothed with fine hairs (cilia), which beat rhythmically thus creating micro-currents which drive small particles, such as diatoms (on which the Cushion starfish feeds, at least in part), towards the mouth.

Common Starlet *Asterina gibbosa*
Closely related to the Duck's Foot starfish is the abundant Common starlet. They are both members of the class *Asteroidea*. No other species of which is known from Irish waters. It is the smallest of the Irish Starfishes, 25mm in diameter being one of the larger specimens. It is pentagonal and very convex – gibbous meaning convex or hump-backed. On the top

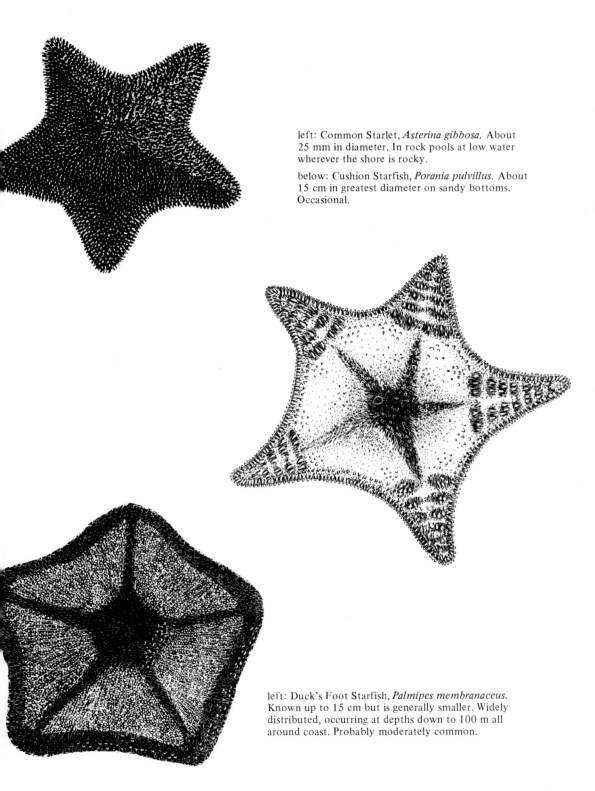

left: Common Starlet, *Asterina gibbosa.* About 25 mm in diameter. In rock pools at low water wherever the shore is rocky.

below: Cushion Starfish, *Porania pulvillus.* About 15 cm in greatest diameter on sandy bottoms. Occasional.

left: Duck's Foot Starfish, *Palmipes membranaceus.* Known up to 15 cm but is generally smaller. Widely distributed, occurring at depths down to 100 m all around coast. Probably moderately common.

surface it is covered with groups of spines in regular rows. Usually it is greenish-yellow in colour but may have a reddish or brownish hue on occasions, and it is to be found wherever the Irish coastline is rocky. It is easily found in pools at low water where it shelters under the rocks. It fastens its eggs to stones and is unusual in that the young larvae do not undergo a pelagic swimming stage, but crawl over the bed of the sea. This appears to be an adaptation to shore life and would prevent the young stages from being swept out to sea.

Duck's Foot Starfish
Palmipes membranaceus

This almost paper-thin, pentagonal-shaped starfish is remarkably like the web-shaped foot of a duck, the five rays having thin, rather leathery webs or lobes stretching between them. The central discs, the rays and the outer border are bright red in colour — the remaining part of the lobes being white. In diameter it has been known to measure up to 15cm, but it is generally much smaller.

The Duck's Foot starfish feeds on crustaceans, such as small crabs, molluscs, and other echinoderms, which it swallows whole. After digesting the soft tissues of its prey, it ejects the hard parts. No starfish species is able to break up or chew the horny, bony or shell parts of marine animals. It is a widely-distributed species, occurring from the Mediterranean to the Arctic Seas, being dredged from moderately deep water — 800 or so metres in depth.

Common Sunstar *Solaster papposus*

This large and handsome species is common around the coasts of Ireland on coarse sand and gravel in depths of 30 to 70m and often strays elsewhere — probably from the refuse of trawlers. The disc is broad, about twice the length of the rays in diameter. There are about 12 to 13 rays, but can number up to as many as 15. The whole upper surface is covered with tubercles from which arise a bundle of long spines, with about five rows of tubercles along each ray.

In colour, the Common sunstar is very

Common Sunstar, *Solaster papposus*. Specimens of 25 cm not uncommon. Generally distributed around the western coasts of Europe.

variable. The whole of its upper surface may be deep purple, although those seen and illustrated by the author had the central part of the disc red, with radiating stripes of red and the rays banded with the same shade. Specimens are known entirely red, except for the tubercles which are green. They usually grow to a large size with 25cm in diameter not uncommon. They feed on bivalve molluscs, pulling them open, but small specimens are eaten whole. It has a general distribution throughout the western coasts of Europe.

Common Starfish *Asterias rubens*

This well-known inhabitant of Irish waters may sometimes be found at the seaward edge of the shore where it shelters under rocks and amongst debris and seaweed, but generally occurs at a depth of from 1 metre to 60m. There are five tapering rays which are more or less pointed at their apices — although the number of rays may occasionally be four or six. In colour it is extremely variable, being from yellow to red to orange or reddish-brown to purple, with the spines and their bases white. Sometimes reaching as much as 25cm across, most specimens found are only half this size.

The Common starfish is an important predator of bivalve molluscs and causes considerable harm to oyster and mussel beds. Additionally, they feed on a large variety of animals which they come into contact with during their rather slow wanderings. Small molluscs are swallowed whole but larger specimens are opened by means of a sustained pull. The starfish wraps itself around both valves of the mollusc, gripping it with its sucker feet. The mollusc tires first and allows its valves to open slightly. The starfish then injects its stomach through the gap and digestion slowly commences. Crabs are sometimes treated in this way. The suggestion, however, has been put forward, that the starfish secretes some poisonous fluid, possibly paralyzing or anaesthetizing the prey, weakening it sufficiently to make it relax its muscles. Many starfishes, the Common starfish included, possesses the ability to regenerate limbs. If the starfish loses one or more limbs, then they grow again and are often found with some of their limbs of normal size, and the remainder small, showing where the old limb broke off by the abrupt change in the thickness of the limbs.

Common Brittlestar
Ophiothrix fragilis

Brittlestars differ from starfishes in having a clear distinction between the central disc and the five flexible, snake-like rays. The latter break off with great ease — hence the popular and specific name! Brittlestars constitute the class *Ophiuroidea* (*ophis* is Greek for snake), and it is a very successful group. Not only are 2,000 or so different species known, but they often occur in extraordinary numbers — over 3,000 individuals of our Common brittlestar per square metre of the sea-floor are recorded. This species is found, again in great numbers, around the European coasts from low-water to about 70m or even deeper. It appears to prefer rocky or coarse gravel bottoms.

The disc is about 1 to 2cm in diameter, with two triangular plates at the point of insertion of each ray — the rays are usually from four to five times the diameter of the disc. In colour it is extremely variable, hardly any two being alike and generally of two definite basic colours such as red and white, green and grey, red and yellow,

and blue and grey; the rays themselves are often banded in two colours. Their exquisiteness has captivated many marine biologists, and stimulated them to try to capture their beauty in colour photographs.

Edible Sea Urchin
Echinus esculentus

Sea urchins are globose, very spiny, echinoderms, showing a five-sided radial symmetry. This is most evident on the sculpturing of the outer skeleton, which consists of a large number of interlocking chalky plates with radiating rows of convex tubercles, being the basal connections for the spines. Many plates are pierced for the tubed feet, and the whole outer skeleton is known as the 'test'. These 'tests' may be found by the casual observer along many rocky coasts, and they are of great beauty, but, it seems pointless to kill these delicate creatures in order to provide a test as a holiday souvenir.

At the centre of the top surface of the test there are ten plates which constitute the apical system. Eggs or sperm are ejected from pores in each of five genital plates. One plate connects with the water vascular system of the animal, and is pierced by a number of small pores and this is known as the madreporite. All these plates are peripheral to the anus, and it is through this that indigestible food material is excreted. At the centre of the bottom surface of the test lies the peristome or mouth, and through this the fine, hard, calcite teeth project, making a complex, five-rayed jaw apparatus of 20 parts — known as the Aristotle's Lantern. There is an extraordinary microstructure of specialized spines which, reluctantly, the author feels is beyond the scope of this present book to describe. If the reader is stimulated to further reading, there are references given elsewhere.

The Edible sea urchin is about 12cm in diameter across the test, and globular but flattened at the bottom. There are many variations in shape, probably due to age. In colour it is generally reddish with white tubercles, and the spines are red and white also, but these are relatively short. Around the mouth spines the polychaet worm, *Scalesetosus assimilis,* often occurs. It was formerly abundant around the

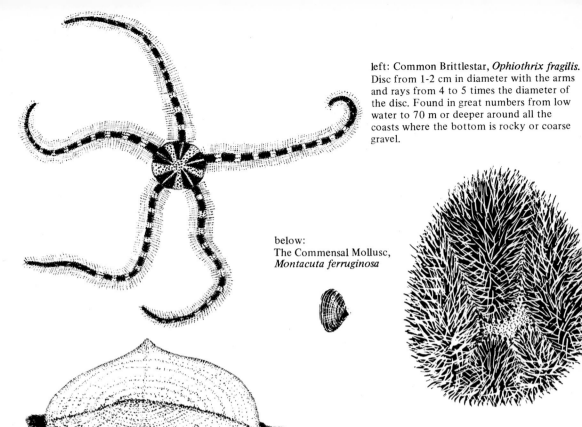

left: Common Brittlestar, *Ophiothrix fragilis.* Disc from 1-2 cm in diameter with the arms and rays from 4 to 5 times the diameter of the disc. Found in great numbers from low water to 70 m or deeper around all the coasts where the bottom is rocky or coarse gravel.

below:
The Commensal Mollusc,
Montacuta ferruginosa

above: Common Heart Urchin, *Echinocardium cordatum.* Maximum length about 6 cm. Occurs in sandy bays all around European coasts, often in great abundance.

below: Purple Sea Urchin, *Paracentrotus lividus.* Excluding the spines the diameter of the body is about 50 mm. Spines average about 25 mm in length. Occurs only around the South-west coast, being most plentiful in Co. Cork.

above: The By-the-wind Sailor, *Velella velella.* 'Skeleton' about 3 cm in length. Often thrown up on to the western and southern beaches in extraordinary numbers.

below: Edible Sea Urchin, *Echinus esculentus.* About 12 cm in diameter. Formerly abundant around the coast to a depth of 100 m but much less common today.

Irish coasts down to a depth of about 100m, but now skin-diving activities have probably been the cause of it being much less common today. It has been collected from as far down as 1,000m.

Purple Sea Urchin
Paracentrotus lividus

This sea urchin, or 'Egg Urchin', as they were once called, is covered with long, slender, purple spines averaging about 25mm in length. Sometimes the spines may be a deep olive-green. The body, without the spines, measures about 50mm. It occurs on the west coast, from Dunmanus Bay to Donegal, and one locality on the south coast — Loch Ine, Co. Cork. It is a Mediterranean species.

In some localities it covers rocks just below the surface at low-tide level and generally prefers crevices and ledges. When living in wave-exposed situations it can excavate a hollow in the rock, and it can generally only be removed with difficulty. The boring seems to be carried out by continual abrasion of the spines around the mouth and the five teeth. A closely-related species, *Strongylocentrotus purpuratus,* found in California, is known to bore a shelter into the steel pillars of harbour works!

The Purple Sea Urchin is, of necessity, gregarious, in that the young require the shelter of their parents' spines in order to reach adulthood. In the cause of conservation it is unwise (to say the least) to strip the rock of this interesting marine animal, purely for the delight of gourmets.

Common Heart Urchin
Echinocardium cordatum

This heart urchin spends its life buried in the sand from between tide-marks to at least 70m, and appears to be often associated with the sand of eel-grass *(Zostera)* beds. Found around the coasts of Europe wherever there are sandy bays, it is generally very common. After storms, when there has been much water turbulence, large numbers are thrown onto the shore where they are blown about by the wind. In spite of their rough treatment the fragile 'test' often remains intact, although the spines are soon lost.

Heart urchins differ from sea urchins in that they show a bilateral symmetry which is superimposed, as it were, on the basic radial plan. There is a furrow on the front margin of the body, giving it a heart-shape when the animal is viewed from above and it is covered completely with a felt of small spines. The Common heart urchin is specially adapted for burrowing in soft sand, into whose depths it can sink — pushing and covering itself completely in about ten minutes. Its ventral spines are modified as paddles for the purpose of pushing the sand aside. It does not possess teeth as do the sea urchins, as its food consists only of detritus. This is 'picked over' by its tube feet, and the mouth is located towards the front end of the body with an anus to the rear. The waste material voided from the anus is less likely to contaminate food intake near the mouth. At its longest axis, the Common heart urchin is about 6cm in length. A commensal (closely associated) bivalve mollusc, *Montacuta ferruginosa,* frequently accompanies it.

Sea Cucumber *Holothuria forskali*

Sea cucumbers classified *Holothuroidia* are elongate, often slimy, echinoderms possessing a ring of tentacles, often much branched, around the mouth. These are part of the complex water vascular system common to all echinoderms, of which the tube feet are components. An anus is at the opposite end to the mouth. The skin is leathery and within it are embedded variously-shaped calcite spicules. Fossil holothurians are known with the calcite spicules arranged much as in the 'test' of sea urchins. Although cylindrical, many sea cucumbers have developed a flattened area or 'sole', on which they creep. They are widely distributed throughout the seas of the world, from the inter-tidal zone to abyssmal depths, and about 900 species are known, with about a dozen of these known from Ireland.

Holothuria forskali is a relatively large, black species found around the Irish coast at a depth of 10m or so. It appears to prefer a rocky sea-bottom with clean gravel immediately adjacent.

The specimen drawn was 15cm long. This species is often known as the 'cotton-spinner' on account of its habit of ejecting from the anus white, cotton-like organs, which are very sticky and which enmesh an attacking predator. Some species of sea cucumbers which do not possess these organs are able to escape predation by eviscerating themselves — leaving their stomachs for the predator to feed upon whilst making an escape. A new stomach will grow later.

COELENTERATA

This large and major group of primitive animals, for convenience, is dealt with next to the echinoderms, although there is little relationship between them. As an example the writer has taken only the skeletal part of *Velella velella* known as the 'By-the-wind-sailor'. The *Coelenterata* are relatively simple bag-like animals. They are built on a circular plan, i.e. there are not two symmetrical sides.

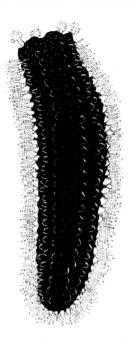

Sea Cucumber, *Holothuria forskali*. About 15 cm in length, but attains much greater size. Found around Irish coast to a depth of about 10 m where the bottom is rocky or gravelly.

The By-the-Wind Sailor
Velella velella

To the class *Hydrozoa* belong a number of small, inconspicuous animals which mostly go unnoticed by non-naturalists. Perhaps the nearest that hydrozoans get to recognition is when they form an encrusting growth on piling and other harbour works, or when they appear as a seaweed-like outgrowth on rocks and stones in the sea. Almost all members of this class of primitive organisms are marine. However, a few are found in freshwater, and the common *Hydra viridis* is well known in ponds throughout Europe. This species is of a simple, bag-like construction with tentacles around the mouth of the 'bag'. Hydrozoans exhibit two structural forms — the polynoid form, which is *Hydra*-like, and the medusoid form which is saucer-like, surrounded by blunt tentacles and with a central tube-like mouth. Some species of *Hydrozoa* pass through both forms in the course of their life-cycle.

Although *Hydra* and its near relations live as solitary polyps, most hydrozoans form colonies. The colony is produced by 'budding', and the new buds persist and eventually produce still more buds, with the result that each polyp is connected to all others in the colony — a hydroid colony. Such colonies are usually found anchored by a horizontal, root-like mat. Many species show di- or poly-morphism. This is where the polyps are modified to perform different functions, such as stinging the prey, trapping it, and passing it to the mouth-like polyps for digestion. Several species are free-floating on the tropical oceans.

The By-the-wind Sailor has a skeletal part consisting of an oval, horizontal plate about 3cm in length, and with an upright 'sail'. The skeletal part is covered with gelantinous tissue — purplish in colour — and hanging downwards are the tentacle-like catching polyps. They transfer the prey to a large, central mouth. In the warmer seas they must be present in countless millions and, when being blown or carried by currents into colder climates, the colony dies; their thin, papery and shining skeletons are thrown onto our southern and western beaches in great numbers.

PART TWO
The Vegetable Kingdom

Introduction to Part Two

When considering the whole of the vegetable kingdom, the flowering plants reach the highest plane of specialization and adaptation. Not only does this relate to their various floral mechanisms, but also to their extraordinary distribution. Except for the most arid, inhospitable regions and the open oceans, they have virtually colonized the Earth. Their importance to mankind cannot be measured. By far the greater part of his food is derived from the various plant organs. Man's other uses of flowering plants are too numerous to mention. In another category, however, it should be stated that their aesthetic value has stimulated art and literature. Indeed, it is an essential part of them.

The flowering plants are technically referred to as the *Angiospermae*. This means that the seeds are produced in a case or outer covering (the ovary), in contrast to the gymnosperms or naked seeded plants, such as the conifers. The ovary supports a tube-like organ — the style — usually at the top, and at the apex of the style is a cushion-like part — the stigma — and it is this which receives the pollen. The latter is produced by stamens, which are the male reproductive organs, and consist of the anther (containing the pollen) and a stalk or 'filament'. The male and female organs may be produced by a single flower or they may be in separate flowers (male and female flowers). Both sexes may be present on a single plant, or they may be segregated onto different plants.

Pollination takes place firstly, by the transference of pollen onto the stigma. This is usually effected by wind or by different flower-visiting insects. There are many special mechanisms which bring this process of fertilization about, but unfortunately descriptions of these are beyond the scope of this very short account. It is interesting to note, however, that in the *Fuchsia* (in its native South America) pollen transference is effected by Humming birds. When the pollen grain comes into contact with the stigma, it germinates by sending out a long, minute tube down the style and into the ovule. The nuclei then fuse with the nucleus of the egg cell, thus ensuring further development of an embryo ovule into a seed, which will eventually germinate to produce a mature plant.

Besides the male and female organs already described, there are parts of the flower which might be described as 'secondary' sexual characters. At the base of the flower there are the sepals — usually four or five, and above them is a ring of petals — of the same number. In some families, however, the sepals and petals may be greatly reduced and, occasionally, increased. One important feature of plants concerns the relative position of the ovary with regard to the points of insertion of the sepals and petals. If the petals are below the ovary, then the ovary is said to be 'superior', but if they arise above the ovary then the ovary is said to be 'inferior'. Dr. D. A. Webb's *An Irish Flora* describes 1,167 species of flowering plants. Of these 964 are given greater prominence than the remaining 203 species, for the latter are rare, only partially naturalized or only identifiable by a specialist. These figures refer to all *Angiospermae*, both Dicotyledons and Monocotyledons.

15 Flowering Plants/Dicotyledons

Lesser Spearwort
Ranunculus flammula
The most primitive, that is, the simplest and least modified of the flowering plants, are thought to be contained in the family RANUNCULACEAE. This is the 'buttercup family', but many other well-known plants belong in addition. It is a large family, containing about 1,300 species in 48 genera — chiefly distributed in temperate and even arctic zones. In Ireland there are 25 species in eight genera.

The Lesser spearwort is a perennial, buttercup-like plant standing up to about 50cm in height, although somewhat shorter generally. The main stem is creeping, and this roots at intervals in the wet, gravelly places in ditches, near lake margins and other similar situations. The upright branches bear spear-shaped leaves whose expanded bases clasp the stem. The flowers of five petals and five sepals reach a diameter of up to 18mm, whilst in the very similar Greater spearwort, *R.lingua,* the flowers are mostly more than 25mm across. Lesser spearwort is abundant throughout the country in damp habitats and flowers in June through to August, but Greater spearwort is found only in central areas and blooms during June and July.

Moorland Crowfoot
Ranunculus omiophyllus
Eight of the 17 Irish species of *Ranunculus* belong to the 'crowfoot' group. Some of these are difficult for the non-botanist to identify — indeed, botanists who are not specialists may not be confident in their identification! All of them are associated with water — from fast-flowing rivers to bogs. All have white flowers with prominent yellow stamens, slender stalks and lobed leaves. In a number of species some or all of the leaves are submerged and finely divided into hair-like segments.

Moorland crowfoot, which has also been named Lenormand's Water crowfoot, does not possess submerged, dissected leaves as it grows on the top of mud with the aerial leaves usually adhering to the mud surface. The three main lobes of the leaves show minor subdivisions and the flowers are up to 12mm in diameter, and they are in bloom from May to August. This plant generally grows in upland, non-calcareous localities, and its distribution abroad is from north-west Britain, through Belgium and France, to northern Spain and Portugal.

Welsh Poppy *Meconopsis cambrica*
The poppy family, the PAPAVERACEAE, is one of the most uncomplicated or 'primitive' families amongst the Dicotyledons. The characters of this family are well exemplified by the Welsh poppy. Throughout the world, but

above: Moorland Crowfoot, *Ranunculus omiophyllus*. Grows flat on mud surfaces in upland, acid areas. Fairly common in the South and South-east.

left: Lesser Spearwort, *Ranunculus flammula*. Height to about 50 cm. Abundant throughout the country in damp situations.

right: Welsh Poppy, *Meconopsis cambrica*. About 30 cm in height. Rare in mountains but may be found near houses as garden escape perpetuating itself.

mainly in the northern hemisphere, and excluding the Arctic and the Tropics, various poppy species are to be found growing wild. However, it is not a large family compared with others, with a world-wide number of 115 species classified in about 25 genera. In the Irish Flora, there are four genera containing eight species.

The familiar Corn poppy or Field poppy, *Papaver rhoeas,* must be one of the best known of plants and a weed of cornfields, formerly much more abundant than it is today. The Welsh poppy is a plant of damp, mountainous areas where the ground is well-drained. It is also found around houses in some districts where it per-petuates itself by seeding down. A hairy perennial about 30cm in height, it produces large, yellow flowers and when injured exudes a yellow latex. The seed capsule is about three times as long as it is wide.

below: Horned Poppy, *Glaucium flavum.* May reach 1 m in height. Common locally but absent from the North-east.

above: Dame's Violet, *Hesperis matronalis.* Usually about 60 cm in height. A European introduction now commonly found in the neighbourhood of old gardens.

below: Marsh Violet, *Viola palustris.* Creeping close to the ground its total height not much more than 10 cm. Abundant in some areas but rare in central region.

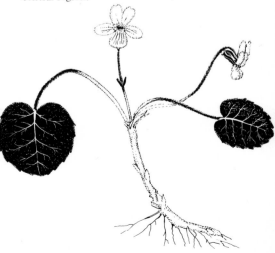

150

Horned Poppy *Glaucium flavum*

Another yellow Irish poppy is the Horned poppy of the sand and shingle beaches. It is a stout, bluish-green plant with hairy, much-divided leaves and is supported by a strong, deep tap-root. It may reach almost 1m in height. If injured, the wounds exude a thickish, yellow latex. The short-stalked flowers are up to 9cm across, and the two hairy sepals fall away when the four petals unfold. Perhaps the most characteristic feature of this plant is the exceptionally long seed capsule — up to 30cm in length! It splits along its length when ripe. The Horned poppy blooms from June to September. There are about 21 species in the genus, mostly Mediterranean in distribution. In Ireland, it is locally frequent, except in the north-west.

Dame's Violet *Hesperis matronalis*

The family CRUCIFERAE, in which this species is classified, has a simple floral organization. Both sepals and petals are in fours and are free — that is, not joined. There are six stamens of which two are generally shorter, although on occasion four only may be present. The superior ovary consists of two fused carpels. A dry fruit is produced, which splits to scatter the seeds. Many familiar vegetables and garden plants belong to this easily-recognized family — cabbage, cauliflower, turnip, radish, wallflower, stock, honesty, candytuft, aubretia and alyssum. Dame's violet is hairy, bearing several deep purple flowers on an erect stem, which may be up to 1m in height, although more usually about 60cm. It is biennial. The flowers are about 18mm in diameter and are fragrant at night. They are attractive to butterflies, moths and the caterpillars of the Orange-tip butterfly, *Euchloë cardamines*. The latter feed on the serrated, spear-shaped leaves. The plant is in bloom from May to August. Introduced from Europe, Dame's violet is now well-known as a garden-escaper, and frequently seeds itself down year after year in neglected places and along hedgerows adjacent to old gardens.

Scurvy-Grass
Cochlearia officinalis

At one time this sharp-tasting biennial or perennial herb was eaten, generally by sailors. On long voyages, during sieges or shipwreck, it was a source of Vitamin C (ascorbic acid), essential for bodily health. Scurvy was thought to be a disease associated with *excessive* consumption of certain foods in a monotonous diet, and it was not until 1753 that it was stated to be associated with a diet *deficiency*.

The fleshy leaves of this plant are variable in shape, being triangular to kidney-shape, with long stalks. The flowers are white, about 12mm in diameter and form a dense raceme, attractive to a number of insects. It may be found in a variety of coastal situations — from salt-marshes to cliffs, around the whole of the Irish shoreline.

There are three Irish species in the genus *Cochlearia* of the CRUCIFERAE, together with a number of hybrids, but only *C.officinalis* is common. The others are rare.

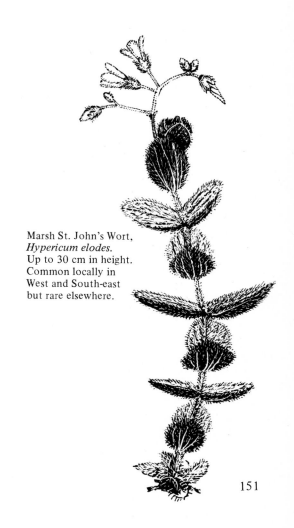

Marsh St. John's Wort, *Hypericum elodes*. Up to 30 cm in height. Common locally in West and South-east but rare elsewhere.

151

Marsh Violet *Viola palustris*

Members of the family VIOLACEAE are quite characteristic, and cannot be mistaken for others. Indeed, they are well-known as violets, pansies and violas, with a number of hybrids popular as garden flowers. There is one genus only in Ireland containing eleven species, although world-wide the family is divided into 16 genera with about 800 species.

This low-growing, herbaceous perennial with a long, creeping rhizome has a short stem, but the long-stalked leaves are round, inclining to a kidney-shape and they are slightly toothed. The rather small, pale-blue flowers or lilac coloured blooms have streaks of a deeper hue. The base of the sepals project beyond the point of attachment, and the lowest of the five petals is produced backwards — as a spur. The Marsh violet may be found in a number of localities where it is predominantly wet — such as in bogs, in marshes and even to the top of the highest mountains where moisture remains. It is common and in some areas abundant, and yet rare in the centre of the country.

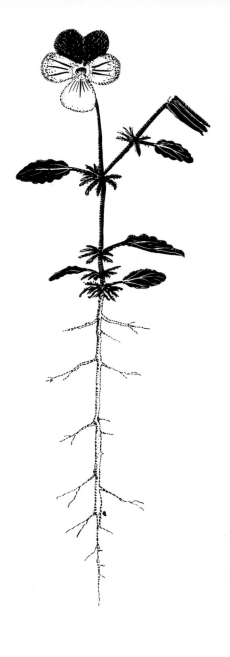

above: Wild Pansy, *Viola tricolor.* Flowers about 18 mm across. Occurs occasionally and mainly in the North and East.

left: Tutsan, *Hypericum androsaemum.* Up to 60 cm in height. Frequently occurs in hedges and sometimes in rocky areas generally throughout the country.

152

Wild Pansy *Viola tricolor*

This species exists in two forms — the sub-species *tricolor*, which is the one illustrated, has large flowers which are about 18mm across, yellow, deep violet-blue or a combination of both colours, and generally the upper petals are darker with dark streaks on the three lower ones. It is an annual and is found chiefly in the north and east. The other form — *curtisii* — is perennial and usually grows around sandhills and shingle near the sea.

Tutsan *Hypericum androsaemum*

There are twelve Irish members of the family GUTTIFERAE, all in the 'St. John's wort' genus, *Hypericum*. Worldwide, it is a large genus with about 220 species — all occurring in temperate regions. All the species have conspicuous, yellow flowers with five petals, five sepals and many stamens. They do not secrete nectar. The leaves are opposite, without stalks and stipules.

Tutsan is a bush with the lower stems of a woody texture. It reaches a height of about 60cm and has a slightly aromatic odour. Its oval leaves are spotted with minute, white glands, and are up to 8cm in length. The flowers are up to 18mm in diameter and are in bloom during July and August. The attractive fruits are at first red but later become a purplish-black colour. This plant is frequently found along hedges, edges of woods and waste, rocky places. Its name is derived from the French 'tout-saine'.

Marsh St. John's Wort *Hypercium elodes*

This is one of the most characteristic of the *Hypericum* genus. The almost round, very woolly leaves clasp the stem and are unwettable. The stems arise from a creeping stolon, often mostly submerged in marshes and acid bogs. The few pale yellow flowers are borne on a drooping spray, and are about 12mm in diameter. The sepals are edged with red glandular teeth which give the flower an orange-tinted appearance. It is in bloom in July and August, and is a common plant in suitable situations in the west, south-west and south-east, but is seldom seen elsewhere in Ireland.

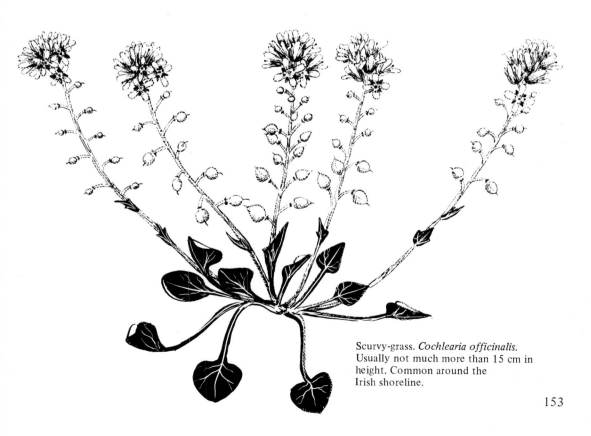

Scurvy-grass. *Cochlearia officinalis.* Usually not much more than 15 cm in height. Common around the Irish shoreline.

153

Trailing St. John's Wort
Hypericum humifusum

The smallest of the Irish St. John's Worts, this perennial forms a rosette of slender, wiry or woody stems which usually lie on the ground. This is a plant, which is always seen from above, and it has been drawn this way. The opposite, elliptical leaves are up to about 1cm in length and the top surface is covered with translucent glands. Its pale yellow flowers are few in number on each stem and are about 1cm across, and may be seen in bloom during July and August along the edges of sandy paths and banks, as well as in peaty areas. It is occasional in Ireland. The specimen illustrated was from Co. Cork.

Greater Stitchwort *Stellaria holostea*

The family CARYOPHYLLACEAE contains 13 genera and 37 species in Ireland. Worldwide there are 70 genera and about 1,450 species, chiefly in the temperate northern hemisphere. The family includes soapwort, *Saponaria officinalis,* often found in abundance in damp situations near houses, but mostly in the southern half of the country. The campions, corn-cockle, and chick-weeds are also related, but corn-cockle is now rare.

The Greater stitchwort is a conspicuous plant in May and early summer along hedges. The stalkless leaves are up to 5cm in length and are in opposite pairs. It is much commoner in the eastern half of the country than in the west where it is local. The specimen illustrated came from Brittas Bay, Co. Wicklow.

Sea Campion *Silene vulgaris*
subspecies *maritima*

Sea campion is a sub-species of *S.vulgaris,* the Bladder campion, although there is some controversy as to its specific rank. The two forms hybridize. This member of the CARYOPHYLLACEAE is found on cliffs, rocky shores and other localities near the sea. It is chiefly distinguished by the large, bladder-like calyx with six recurved teeth. The white flowers are from 20 to 25mm across and are deeply divided. There are only two to three on a flowering shoot. The plant stands up to 25cm

in height but in exposed situations it may only reach 10cm. The flowering shoots are erect with long internodes and are often angled at the nodes, whilst the non-flowering shoots are prostrate.

Ragged Robin *Lychnis flos-cuculi*

Another member of the family CARYO-PHYLLACEAE deserves a mention when dealing with Irish plants. This is Ragged Robin which often grows so profusely in marshy fields and along damp hedges. The large patches of pink are very arresting and quite unlike any other Irish plant. It grows to a height of about 60cm, and the stem and leaves are slightly hairy. Only the lower leaves are borne on stalks. The colour of the flower is reddish pink and each of the five petals is divided into four narrow lobes. The flowers are to be seen in June and July.

Greater Stitchwort, *Stellaria holostea.* Stems may be up to 60 cm in length. Commoner in the eastern half and local in the West.

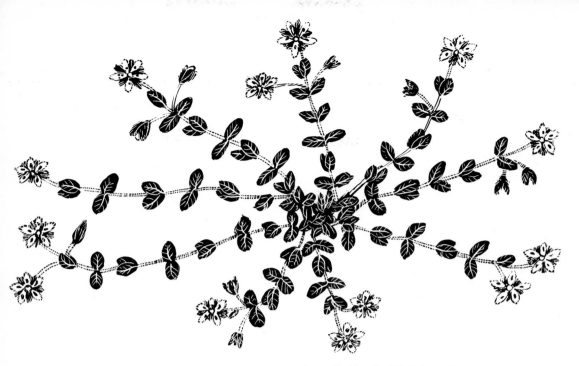

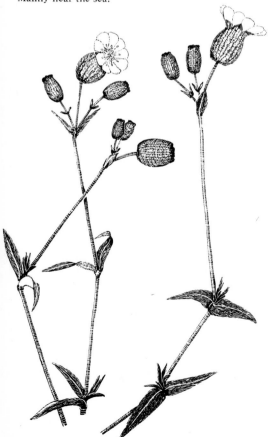

Sea Campion, *Silene vulgaris ssp maritima.* Up to 25 cm in height but may be only 10 cm. Mainly near the sea.

above: Trailing St. John's Wort, *Hypericum humifusum.* Trailing stems up to 25 cm in length. Occurs only occasionally, usually along sandy paths and in peaty situations.

Tree Mallow *Lavatera arborea*

Widely distributed throughout temperate and tropical regions, the MALVACEAE contains about 900 species in 40 genera, and includes such well-known plants as the temperate hollyhock and the tropical *Hibiscus,* together with the economically important *Gossypium,* from which cotton is spun from the seed-coat hairs. Ireland possesses three genera containing five species all known as Mallows.

Tree mallow is a stout, erect biennial, but sometimes perennial. The lower stem may be woody, and the plant occasionally reaches a height of 3m. The large leaves are lobed more or less acutely, and are woolly. Its rose-purple coloured flowers are veined in deep purple and are up to 4cm in diameter. There are three large bracts beneath the calyx. The numerous anthers form a tube or column through which the stigmas project. This handsome plant occurs on rocks and waste ground close to the sea and frequently close to houses, but is fairly rare, being found only on the east and south coasts and on the west coast as far north as Co. Clare.

Wood Sorrel *Oxalis acetosella*

This is the sole Irish representative of the family OXALIDACEAE, which consists of eight genera with almost 900 species. It is a delicate-looking plant identified by its three heart-shaped leaflets, borne on a slender stalk arising from a creeping perennial rootstock. The leaves have an acid taste from which the common and the scientific names are derived. The solitary flowers, which are up to 18mm across, are borne on long stalks and are white with purple veins. Sepals, ovary cells and petals are all five in number, but there are ten stamens of which five are short and five long. It is common practically everywhere there are shady woods and hedges.

Field Maple *Acer campestre*

Two members of the ACERACEAE family are to be found wild in Ireland, but neither of them is considered to be native. The well-known sycamore, *Acer pseudoplatanus,* which grows to a fine tree with grey bark and five-pointed leaves which are large and toothed, is common almost everywhere in hedges, woods, parks and in gardens. It is a most prolific tree because of its self-sown, two-winged fruits.

The Field maple or English maple, *Acer campestre,* has been planted especially in hedges, where it usually grows well. In fact in the Irish climate it occasionally does too well and runs wild! In form it is a bush or small tree, and if clipped when growing in a hedge will become dense. It is easily identified from the sycamore by its leaves which are much smaller, with blunt lobes and no serration. In addition, the two wings of the fruit are in the same straight line, whereas in the sycamore, these diverge at right-angles or thereabouts.

Burnet Rose
Rosa pimpinellifolia

This species, also called Scotch rose, is among the 2,000 or so species in the 90 genera of the typically temperate family ROSACEAE. Seventeen of the genera occur in Ireland. In the genus, *Rosa,* there are 13 Irish species but some are difficult to determine, and the situation is further complicated by intermediates and

hybridization. Burnet rose however, is easily identified as it is an erect shrub, about 30cm in height and the stems are completely covered with numerous, straight thorns of varying lengths. The leaves consist of seven- or nine-toothed leaflets and there are fairly prominent stipules, as in all ROSACEAE. The flower is a single rose of typical appearance, white or pink, with five

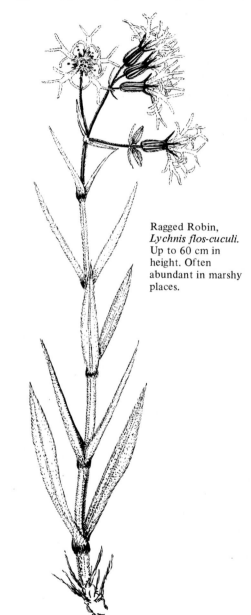

Ragged Robin, *Lychnis flos-cuculi.* Up to 60 cm in height. Often abundant in marshy places.

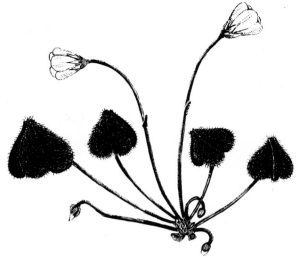

above: Wood Sorrel, *Oxalis acetosella*. Usually no more than about 10 cm in height. Common where there are shady places.

below: Field Maple or English Maple, *Acer campestre*. Up to 15 m in height. Introduced and often planted in hedges.

above: Tree Mallow, *Lavatera arborea*. Occasionally attains a height of 3 m. Rare, on sea-cliffs in the South and East. Also near houses close to the coast as an escape.

right: Burnet Rose, *Rosa pimpinellifolia*. About 30 cm in height. Generally common near the sea on sandhills and rocky heathland.

157

wide petals with their outer margin concave, and the styles hardly projecting from the numerous stamens. The fruit is usually black with persistant calyx teeth. Flowering in May and June, and possessing a strong fragrance, it is generally common near the sea on rocky heathland and sandhills, frequently being covered or almost smothered by sand; in other situations it is rare.

Shrubby Cinquefoil *Potentilla fruticosa*
The genus *Potentilla* contains over 300 species, mainly temperate in distribution, and Ireland has seven of these. The Shrubby cinquefoil is a well-known garden shrub which does well in basic soils and is a native of Clare, Galway and Mayo. It grows to about 60cm in height and stems and leaves are more or less covered with soft hairs. The 'bark' peels off the woody stems when about three years old. Each leaf consists of five leaflets. The flowers are yellow, from 18 to 36mm in diameter and are generally unisexual. This attractive plant occurs in rocky areas which flood occasionally, and is in bloom from June to the end of July.

Stone Bramble *Rubus saxatilis*
This locally distributed bramble is found where it is stony or rocky and usually in hilly regions. It generally grows where the soil is basic, and perhaps nowhere is it to be found so abundant as in the crevices of the limestone pavement of the Burren in Co. Clare. Most visitors from Britain are more likely to find it for the first time in Ireland long before they discover it in England, Scotland or Wales. In a genus in which well over a hundred species are listed for the British Isles and over four hundred for the North Temperate zone, it might be thought a difficult species to identify, but happily this is not so.

Mountain Avens *Dryas octopetala*
One of the most extraordinary botanical sights in Ireland must be the 'Mountain' avens in bloom around the Burren in Co. Clare in May, June and July. Praeger states, 'the thousands of acres of this arctic-alpine plant is one of the loveliest sights that Ireland has to offer the botanist'.

This very beautiful 'creeping undershrub' is alpine throughout practically the whole of its extensive range but, in the Burren of Co. Clare, it descends to sea-level onto the coast, often within a few metres of the sea. It seems a contradiction that although the alpine flora of Ireland is generally poor in species, yet sometimes such species are abundant in anything other than in an alpine environment!

As well as in Co. Clare, it occurs in the area of Ben Bulben in Sligo and to the north, and in much smaller localities in the counties of Donegal (where it is found as high as 433m), Kerry, Antrim and in west Galway, where, as in Co. Clare, it grows at sea-level. In Britain it is local, occurring (with the exception of Sutherland) only amongst rocks in mountainous regions, even up to an altitude of 930m. It was formerly found in north Wales and the Lake District, but is probably no longer growing there. Indeed, the only definite English station is in West Yorkshire. In Scotland it occurs north of Perth and in the Orkneys. Elsewhere, in Europe, it is Arctic and Subarctic in distribution, as well as being found in many high mountainous regions as far south as the Pyrenees, the Apennines and Macedonia. In North America it is an inhabitant of the Rocky Mountains, extending as far south as Colorado.

Meadow Saxifrage *Saxifraga granulata*
Ireland possesses ten species of saxifrage, although some are very rare and local whilst others offer difficulties on account of hybridization. These, together with the Golden saxifrage, *Chrysosplenium oppositifolium,* the currants and gooseberry in the genus *Ribes,* as well as *Escallonia macrantha* (all of which have become naturalized) constitute the family SAXIFRAGACEAE. Most plants are to be found only in the west, north and south of Ireland, but the Meadow saxifrage occurs only near the east coast in pastures and on sandhills. Unfortunately it is very rare. It grows as a slender, erect perennial, usually about 25cm in height. The leaves nearest the ground are kidney-shaped and have long stalks, whereas the upper leaves are generally lobed and stalkless. The flowers are white and up to 2.5cm across, and appear in May and June.

above: Stone Bramble, *Rubus saxatilis.* Stems may be up to 60 cm in length. Occurs locally in stony places. Abundant in the Burren, Co. Clare.

below: Mountain Avens, *Dryas octopetala.* Creeping habit with flowers and stalk up to 8 cm in length. Abundant dow to sea-level in Co. Clare and occasionally found on limestone northwards to Fermanagh.

right: Shrubby Cinquefoil, *Potentilla fruticosa.* Up to about 80 cm in height. Only in Counties Clare, Galway and Mayo.

Another species of saxifrage, *S.hartii,* is known only from the sea-cliffs of Arranmore Island in Donegal. It is found in the high mountains of Arctic Europe, Asia and North America, but rarely in North Wales and Scotland. In many ways it is similar to the Tufted saxifrage, *S.caespitosa.*

left: *Escallonia macrantha.* Up to 4 m in height. Planted as a hedge near the sea but naturalised in Kerry.

below: Meadow Saxifrage, *Saxifraga granulata.* About 25 cm in height. Only near the East coast. Rare.

Escallonia macrantha

From the island of Chiloe off the coast of Chile comes this evergreen shrub which makes an excellent hedge in coastal districts as it is salt-resistant. It has become naturalized in Kerry from distributed seed. The flowers, about 15mm in length, are bright pink as are the tubular flower-buds, and the toothed leaves are dark-green above and light-green underneath, and when crushed, are resinous.

Kidney Saxifrage *Saxifraga hirsuta*

This saxifrage is found in acid conditions amongst shady rocks in Kerry and West Cork and nowhere else in Ireland or Britain. It is fairly local in distribution but is often common where it occurs. It grows up to 1,000m. Only in northern Spain and in the Pyrenees is it native outside Ireland. The leaves are variable in shape — from kidney to crenulate, with from six to thirteen obtuse or apiculate teeth on each side, and with each side covered with long hairs. The flowers have oblong sepals which are also obtuse. The elliptical, 4mm long petals are white with a basal yellow blotch, sometimes with crimson markings.

160

right: St. Patrick's Cabbage, *Saxifraga spathularis.* 15-40 cm in height. Known wild only from Ireland where it has a wide but local distribution in Counties Cork, Kerry, S. Tipperary, Galway, Mayo, Donegal and Wicklow.

left: Kidney Saxifrage, *Saxifraga hirsuta.* Up to 40 cm in height. Only in Kerry and West Cork but common where it occurs.

161

St. Patrick's Cabbage *or* London Pride
Saxifraga spathularis

The saxifrages constitute a very attractive group of plants. To all with any knowledge of flowering plants they conjure up a picture of delicate flowers poised on long stalks. These sprout from a tufted plant seemingly growing straight from some cleft in the rocks – of course it is an Alpine situation. *Saxifraga* literally means 'splitter or breaker of rocks'. Throughout my youth, then into middle-age, I had always assumed that 'London Pride' was, indeed, the selfsame plant as found in south-western Ireland as a native plant. But, alack, as I approach old-age, I find my idols have feet of clay! Saint Patrick's cabbage as growing wild in south-western Ireland is now given the specific name of *spathularis,* whereas 'London Pride' is now thought to be a hybrid between *Saxifraga spathularis* and *S.umbrosa,.* Although London Pride sometimes escapes from cultivation and becomes naturalized, it is, in fact, not known to be truly wild.

S. umbrosa is a native of the Pyrenees, but has been found in Heseldon and Linn Gills in Yorkshire since 1792. It is presumed that it was introduced to these areas from a Pyrenean locality. To confuse still further, the nomenclature of St. Patrick's Cabbage and forms which resemble it closely, some plants are thought to be hybrids between *S.hirsuta* and *S.spathularis.* The characters of both parents are combined in a number of ways, and the plants occur in mixed populations with the parents in various proportions. It is strange that in Galway and Mayo the hybrids occur, although the one parent species *S.hirsuta* is absent. *S.spathularis* is absent from England, Wales and Scotland, but has a relatively wide, but local, distribution in Ireland – occurring in Cork, Kerry, Waterford, S.Tipperary, Galway, Mayo, Donegal and Wicklow. In some localities it grows abundantly and is known up to an altitude of 1,130m.

A description of St. Patrick's cabbage would seem to be almost superfluous, so well known is the plant under the name of London Pride in so many gardens in Britain. One reason for this is the way in which it flourishes even in the most grimy of back-gardens in industrial areas. The leaves vary from spatula-like to kidney in shape, and there may be up to seven acute 'teeth' on each side, with an apical tooth which is at least equal in length to the lateral teeth and may be even longer. The stalk is much longer than the blade of the leaf. The flowers are borne on a long stalk and the white petals are 4-5mm in length with one to three yellow spots at the base and numerous crimson spots on the apex. Due to high rainfall in a sheltered situation *S.spathularis* is found as an epiphyte growing on the boughs of trees 7-10m above the ground (at the Blackwater near Kenmare). The ability of this plant to grow in dry, almost soil-less chinks in limestone rocks has also been noted.

Pennywort *Ubilicus rupestris*

The family, CRASSULACEAE, contains three Irish genera. Pennywort is the sole member of its genus and Roseroot, *Rhodiala rosea,* similarly has no other species in its genus. *Sedum,* however, contains eight species. Only three of the latter are other than rare. Pennywort occurs mostly in the west and south where it grows from rock crevices and walls where acid, damp but well-drained conditions prevail. This fleshy plant may be up to 40cm in height, but is usually only half this size. The glossy leaves are about 'penny' size, our 'old' penny that is, and have a sinuate margin tending to have rounded teeth. The stalk joins the leaf near the centre, and the upper surface of the leaf is concave. This gives rise to one of its common names – Navel wort, as well as to its generic name. Numerous bell-like, greenish-white flowers droop from the thick, fleshy stalk, and are about 8mm in length. The calyx has five long, pointed lobes. There are ten stamens and five carpels, and the plant is in flower from June to August and is easily identified.

English Stonecrop *Sedum anglicum*

The family CRASSULACEAE, and particularly the genus *Sedum,* consists of plants with thick, undivided, fleshy, 'succulent' leaves. English stonecrop is a small, creeping, perennial, whose slender, rooting stems often form mats in arid situations such as tops of walls and rock faces where little soil is present. Tiny branches are produced – a few centimetres in height and some

above: English Stonecrop, *Sedum anglicum.* About 5 cm in height. Found on rocks, dry walls and banks near the sea or in mountains. Rare elsewhere.

below: *Fuchsia magellanica* and *gracilis.* Up to 3 m in height. Introduced and used as a hedge in the West but has sometimes run wild.

above: Common Sundew, *Drosera rotundifolia.* About 10 cm in height. Common everywhere in bogs.

below: New Zealand Willow-herb, *Epilobium nerterioides.* A creeping plant with erect flower stems only a few cm in height. Recently introduced but spreading rapidly. Locally abundant mostly in mountainous areas.

163

of these produce flowers. The leaves, which are often reddish in colour, are from 3 to 5mm in length and clasp the stem, although they are without stipules there is a small spur at the base. The flowers, which are produced in small groups, have five sepals, five petals and ten stamens, and are between 10 and 12mm in diameter. The petals are white with a pinkish colour on the back, and they are in bloom from June to August. This plant is most often found close to the sea or in mountainous regions ascending to 1,000m. It is rarely found in the central regions.

Common Sundew *Drosera rotundifolia*

The DROSERACEAE family, to which the three species of sundew belong, is a very highly specialized group of plants. All have leaves furnished with long, glandular hairs, by means of which insects are trapped and digested. All occur in acid soil conditions and the family is distributed throughout the world, but there are only four genera containing about 100 species.

The Common sundew has more or less rounded leaves which lie flat in a rosette, and they are distinctly red and about 8 to 12mm across. The long stalks are from 25 to 40mm across. The small, white flowers, about 6 or 7mm across, are borne on a long, slender stalk approximately 10cm in length which arises from the centre of the leaf rosette. Blooming in July and August, this plant is common almost everywhere where boggy conditions prevail. The other Irish species of this genus are identified by their leaf length and the shape of the blade. *D.anglica*, has much longer leaves than *D.rotundifolia* and the leaf blade is oblong. *D. intermedia*, has leaves of the same length as *D.rotundifolia* but the blade is narrow and elliptical.

New Zealand Willow-Herb
Epilobium nerterioides

The family ONAGRACEAE contains about 470 species classified into 38 genera and distributed throughout the world — although mostly found in temperate and subtropical regions. A number of species are cultivated as garden plants, such as *Godetia, Clarkia, Oenothera* (Evening Primrose)

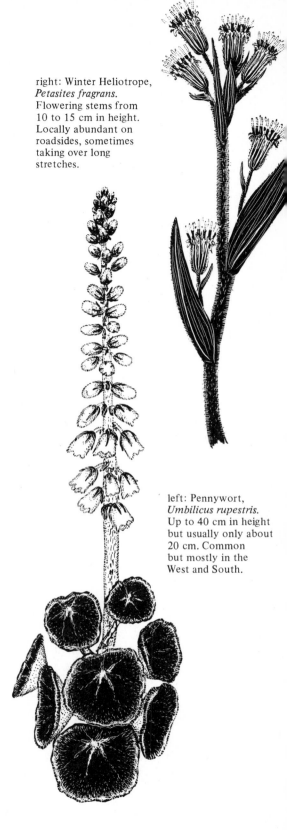

right: Winter Heliotrope, *Petasites fragrans.* Flowering stems from 10 to 15 cm in height. Locally abundant on roadsides, sometimes taking over long stretches.

left: Pennywort, *Umbilicus rupestris.* Up to 40 cm in height but usually only about 20 cm. Common but mostly in the West and South.

and *Fuchsia*. In Ireland, three genera, *Fuchsia, Circaea* and *Epilobium* occur, ten species being known of the last-named. These are the willow-herbs.

New Zealand willow-herb is a very delicate, minute member of the genus, introduced from New Zealand relatively recently — Floras of 1958 do not give it a mention. It is spreading rapidly, especially in mountainous regions where it prefers stony, gritty, well-drained situations. The stems creep over the ground and the round, very slightly notched leaves are produced in pairs, and are from 6 to 12mm in length. The single, white or pinkish flowers are borne on upright stalks, and the fruit splits into four from the top, as in the other members of the genus. The specimen illustrated was collected in Co. Kerry.

Fuchsia magellanica *or* gracilis

One of the most interesting features of the hedges in the west is the exuberance of the *Fuchsia*. The pendulous flowers of scarlet and purple make a fine show of colour from the very end of May until September. The leaves are more or less toothed and have a short, red stalk. The stems of the new shoots are also rich-red in colour, and there is an abundance of these when the plant is severely pruned (like *Escallonia*), making a fairly dense hedge. There are two species involved — *Fuchsia magellanica*, which, apparently, originated in the Falkland Islands and has the purple inner-tube of the flower less than half as long as the narrow sepals, and *Fuchsia gracilis*, from Mexico, whose flower inner-tube is more than half as long as the narrow sepals.

Sea Holly *Eryngium maritimum*

The UMBELLIFERAE, the family to which Sea holly belongs, is one of the most familiar and generally one of the most easily identified of all flowering plant families. The inflorescence is an umbel, from which the family takes its name. This is where the flower stalks radiate from a single point. Most umbels are 'compound', in which the stalks of a number of partial umbels meet at a point to form the general umbel. Other characters of the UMBELLIFERAE are the much-divided leaves (but not shown in Sea holly), a swollen sheath forming the base of the petiole and furrowed stems. It is a very large family, with about 200 genera in which there are about 2,700 species. In Ireland there are 29 genera containing 42 species.

Everyone should be able to identify Sea holly, even from a distance. The powdery bluish-green colour; the very long and strong spines on the white-veined leaves; the dense prickly heads of pale blue flowers, all make this plant unmistakable. It is a rigid perennial, seldom exceeding 60cm in height, and in exposed situations, is

Sea Holly, *Eryngium maritimum.* Attains about 60 cm in height. Generally distributed in sandy, coastal areas.

often much smaller. Sea holly blooms in August and September, and although the colonies it forms are rather local, it is generally distributed around the coast, on sand dunes and in other sandy areas.

Winter Heliotrope *Petasites fragrans*

The COMPOSITAE is the largest family of flowering plants, comprising over 900 genera with about 14,000 world species. In Ireland 84 species occur in 55 genera, and this does not include the fifty 'species' of hawkweed which are identified only by specialists. The genus *Petasites* contains two species which occur in Ireland, and both are to be found along damp hedge-bottoms or wet road verges. The motorist often observes them easily identifying them by their leaves. These are the Winter heliotrope and butterbur, *Petasites hybridus*.

The inflorescence of the former carries up to ten thistle-like heads of a rather pale, watery-looking lilac colour. But they are not nearly so robust as those of Butterbur, but nevertheless are reminiscent of them. However, Winter heliotrope possesses a strong vanilla-like scent which butterbur does not have; the leaves of the former are nearly round, toothed and green on both sides, and these persist throughout the winter — an easily recognizable feature but strange in a plant whose origins are the western Mediterranean!

Winter heliotrope is local, but often exceedingly prolific where it does occur. Driving out from Bantry on the Cork City road it may be observed on the left road edge when in bloom from December to March, when few other plants are in flower. Pull the car window down and smell it. . .

Butterbur *Petasites hybridus*

Contrasting with the Winter heliotrope, the leaves of butterbur die away at the year's end, creating a large, bare area where once the plant flourished. Nothing else can grow where such large, angled, and white-underneath leaves have spread. They are larger than the leaves (otherwise similar) of coltsfoot, *Tussilago farfara*. Springing from the bare earth during April to May, the

Mountain Cudweed, *Antennaria dioica*. From 5-20 cm in height. Rare except for the West and the Centre.

Butterbur, *Petasites hybridus*. Flowering stems from 10-40 cm in height. Locally abundant in damp places. Female plants rare.

166

large blooms each contain a dense mass of thistle-like, scentless inflorescences of a dull, purple colour. Male and female flowers are carried on distinct plants, although a few male florets are sometimes present in the middle of female flower-heads, and a small number of female florets sometimes occur around the margin of male heads. In Ireland, male plants predominate — female plants being very rare. In older times the butterbur had a great reputation for its use as medicine against fevers, and was a 'great strengthener of the heart and a cheerer of the spirits'.

Mountain Cudweed *Antennaria dioica*

Also known as Mountain everlasting and Cat's foot, this species is a member of the COMPOSITAE. It is a woody perennial and forms small mats, with its creeping stems rooting at the nodes. The leaves are white and silky beneath but a dull green colour on the upper surface. The small group of flower-heads is borne on an erect stem from 5 to about 20cm in height. The tubular florets are either pink or white and the female florets are very slender. Mountain cudweed is a plant of acid, short-turfed mountain pastures, dry heaths, and sandhills. Apart from the west and the centre of Ireland where it is abundant, it is comparatively rare. Flowering in May, June and July, the seed, although fertile, is not produced from a sexual fusion.

Sea Aster *Aster tripolium*

A hairless, rather fleshy perennial with long, narrow, alternate leaves, this plant is another member of the family COMPOSITAE. Its height varies with the degree of exposure and may be between 20 and 100cm high. The flower-heads, which have a 'Michaelmas daisy' appearance, are up to 18mm across and their stalks vary in order that these heads are in a compact mass. The outer strap-shaped florets are purple and the inner ones tubular and yellow. Sea aster is a common coastal plant, usually occurring abundantly where wave action is minimal — such as around salt-marshes and in estuaries.

above: Sea Aster, *Aster tripolium*. From 20-100 cm in height. Common in salt-marshes and estuaries.

right: Mexican Fleabane, *Erigeron mucronatus*. Reaching about 10 cm in height. Only at Lismore, Co. Waterford on walls. Introduced.

Mexican Fleabane *Erigeron mucronatus*

In discovering this little daisy-like plant on the walls of the bridge over the Blackwater at Lismore, Co. Waterford, the author was reminded that our flora is not a static entity. Always it is changing. Rare plants are sometimes lost through climatic changes or human pressures, but the niches vacated by them when they die out are often colonized by plants brought in by other agencies – human beings for example!

It is believed that the plant specimen of the illustration was the *first* to be found in Ireland! It has been known to grow on the walls of St. Peter Port, Guernsey, for over one hundred years and is also deemed to be well established in South-west England. It is, however, a native of Mexico! The flower-heads are remarkably daisy-like, and there are two rows of outer florets which are white above and purplish beneath. Instead of the rosette of thick leaves, however, Mexican fleabane has a branched, leafy stem, which is slightly hairy, with three-lobed leaves which may be coarsely toothed at the apex. It was in bloom in the middle of June.

Sheep's Bit Scabious
Jasione montana

Throughout the world there are about 700 species in the family CAMPANULACEAE which are classified in 35 genera, and in Ireland there are six species in four genera. The best known examples are perhaps the *Campanula* species, such as the Harebell, *Campanula rotundifolia,* which is blue in colour with bell-shaped flowers, and Sheep's Bit scabious, but the latter is not typical of this species – for the numerous, bright blue flowers are very small and are arranged on a disc-like head resembling members of the COMPOSITAE. They differ from them, however, in that the calyx is five-toothed, the florets have stalks and each of the latter produces a capsule containing several seeds. Sheep's Bit scabious is not strictly a scabious as these are classified in the genera *Scabiosa, Knautia* and *Succisa,* which are in the family DIPSACACEAE. This attractive plant occurs in rocky, heathy and sandy places – usually in mountainous areas and is often locally common, but it seldom appears in the central areas of the country.

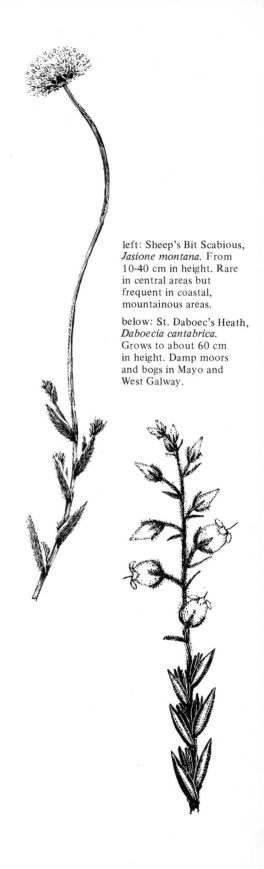

left: Sheep's Bit Scabious, *Jasione montana.* From 10-40 cm in height. Rare in central areas but frequent in coastal, mountainous areas.

below: St. Daboec's Heath, *Daboecia cantabrica.* Grows to about 60 cm in height. Damp moors and bogs in Mayo and West Galway.

168

Irish Heath, *Erica hibernica.* Height about 50 cm to 2 m. Only found in Mayo and Galway.

and rare. Members of the family are chiefly characterized by their simple leaves without stipules, the waxy flowers with tube-shaped corolla consisting of four or five petals with twice as many stamens and a single stigma. Most occur on acid soils (but not *Arbutus*!).

This member of the ERICACEAE is known from the damper heathland and moors of Mayo and west Galway. Elsewhere it is found only in certain localities in Spain. Thriving in light, peaty soil where drought conditions never prevail, and not usually found under tree-cover, this plant grows best, nevertheless, where there is some degree of shade. St. Daboec's heath grows to a height of about 60cm and, although having a heath-like appearance, the leaves are a dark, glossy-green and white underneath, as well as being longer and broader than the typical *Erica* leaves. The egg-shaped flowers are borne in profusion on long, tapering stems and are to be seen from June to October, or even later.

Ivy-Leaved Bellflower
Wahlenbergia hederacea

This rare, extremely delicate perennial is generally found in the south of the country and it is closely related to the genus *Campanula*, but has weak, prostrate stems contrasting with the upright stems of the latter genus. Although the stem may be as long as 30cm, the pale blue, typically long-stalked, campanulate flowers are often difficult to detect amongst the mosses and grasses at the edge of wet areas where it grows. The nearly-round, lobed leaves are borne on long stalks, and the plant is in flower during July and August. The specimen illustrated was found high up on a mountain slope of Macgilly-cuddy's Reeks, but when we returned to base we found that our car had been parked upon it!

St. Daboec's Heath
Daboecia cantabrica

The ERICACEAE is an important family in Arctic, temperate and mountainous tropical regions. There are about 1,500 species in approximately eighty genera. The Irish Flora consists of nine genera containing sixteen species although some of the latter are extremely local

Ivy-leaved Bellflower, *Wahlenbergia hederacea.* Stem may be as long as 30 cm. Rare. Mostly in the South.

169

Mackay's Heath
Erica tetralix mackaiana

Mackay's Heath, as a wild plant, is confined to the wet moors of Galway and Donegal, and found elsewhere in Asturia and Castile in Spain. It was first discovered by William M'Alla in Craiggamore and he passed it on to Mr. J. T. Mackay who was the Curator of the Botanic Gardens (Trinity College) at Ballsbridge, Dublin. It does not set its own seed but the pollen is able to fertilize the closely related *Erica species,* so that it produces hybrids. A very short plant, it grows only to 15cm in height and is prostrate in form. The leaves are shorter, broader and smoother as well as a darker green than those of *Erica tetralix.* The rich pink flowers face all around the stem, and bloom from July to September.

Mediterranean Heath *Erica erigena*
(formerly known as *E.hibernica* and
E.mediterranea)

The naming of this heath is very confusing to the non-botanist. It occurs rather locally on wet moors in Mayo and Galway, and is often found growing in boggy conditions on the shores of loughs — salt or freshwater. Elsewhere it occurs in southern France and Spain. Often flowering as early as January and through to May, the plant is bushy and upright and varies in height from 50cm to as much as 2m.

above: Strawberry Tree, *Arbutus unedo.* From 10-13 m in height. Frequent near Killarney but rare in a few localities in the South-west. Only planted elsewhere.

below: Thrift, *Armeria maritima.* Up to 20 cm in height. A common plant of the sea coasts.

The Strawberry Tree *Arbutus unedo*

If there is one tree above all others that is called to mind when talking or writing of trees special in Ireland, then it is the Strawberry tree, said by Praeger to be the most striking and handsome of the Hiberno-Lusitanian group. It forms trees of from 10 to 13m in height in the woods and on the islands around Killarney and at Lough Gill, but when growing in more or less isolated conditions (at least not in woodland) it is rounded in form and grows up to 7m in height. There are old records of boles 4 to 5m in circumference. Often, however, the branches arise from near the base and a wide-topped shrub results. It is a member of the ERICACEAE (Heaths, Rhododendron etc.) but unlike the

other species in this group it is lime-tolerant — indeed, it is often found growing from crevices in such rocks.

The smooth, dark green and shining leaves are paler underneath, serrated and have a 6mm stalk. In shape they are narrowly oval, tapering towards both ends and 2.4cm in length. The bark is roughish, thin and reddish-brown in colour. The 5cm drooping panicles of flowers are produced later in the year, from October to December, and resemble bunches of Lily of the Valley. The pitcher-shaped blooms are white or pink-tinged and are about 7mm in length. The fruit is strawberry-like, but orange-red and about 18mm in width. Nearly a year is taken for the fruit to ripen so that the previous year's fruit and the new flowers are found together.

The Strawberry tree is native only in Kerry, W. Cork and Sligo, and it occurs throughout the Mediterranean region, extending in France as far north as Brittany. It is not a native of England, Scotland or Wales. It plays a pioneering role in its habitat, generally appearing to originate in a rocky crevice. It survives through the establishment of heaths and through the scrubby oak period, but eventually dies under the oak canopy. The Strawberry tree was formerly much more widespread than at the present time in Irish habitats, but was, in the past, much used for firewood!

above: Spring Gentian, *Gentiana verna*. From 2-6 cm in height. Locally abundant in rocky pastures from mid-Clare to South-east Mayo. Unknown elsewhere.

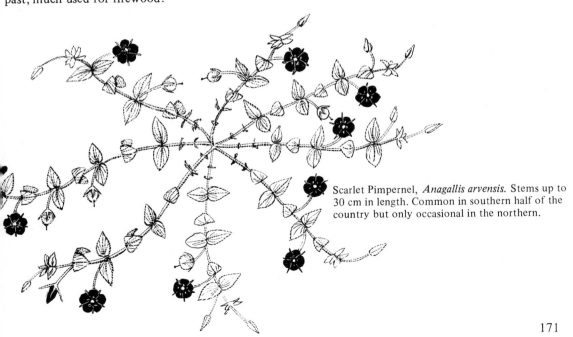

Scarlet Pimpernel, *Anagallis arvensis*. Stems up to 30 cm in length. Common in southern half of the country but only occasional in the northern.

171

Thrift *Armeria maritima*

The family PLUMBAGINACEAE consists of perennial herbs with simple leaves, and with numerous, regular flowers arranged in heads or panicles. There are two Irish genera, *Limonium* containing two species — Sea lavenders, and *Armeria* which has one species.

Thrift, or Sea pink, is a common plant of the sea-coast, only rarely being found elsewhere. The dense tufts of narrow, almost needle-like leaves, generally with only a single vein, and the compact heads of bright pink flowers — often reached by sea-spray, are familiar to almost everyone who has walked along the coastline. The flower-head is encircled by brownish, membraneous scales, as is the inflorescence stalk. There is a basal stock which, in older plants, becomes branched and woody. Thrift has an extensive distribution in the north and is represented by sub-species in Iceland and South Greenland.

Scarlet Pimpernel *Anagallis arvensis*

The PRIMULACEAE is a family of about thirty genera and 350 species, and seven and twelve respectively occur in Ireland. The regular flowers possess a bell-shaped corolla of five segments and a single style and stigma, the stamens adhere to the corolla wall. The primrose, *Primula vulgaris,* and the cowslip, *Primula veris,* are the best known examples.

The Scarlet pimpernel could not possibly be taken for any other species with its scarlet, 8mm star-shaped flowers (open only when the weather is bright), on slender stalks and arising from square-sectioned, trailing stems. It is common in the southern half of the country, but only occasionally occurring in the north. It appears as a weed in tilled fields, and is often abundant in dry, sandy places — on sandhills especially. The specimen illustrated was from Rosscarbery Sands, Co. Cork.

Yellow Pimpernel *Lysimachia nemorum*

The Yellow pimpernel is very frequent almost throughout Ireland, wherever there are woods, damp hedges and mountain pasture. It is a creeping plant, and the long, slender flower-stalks enable the flowers to reach the light above the herbage. The flowers are yellow and the leaves possess only very short stalks.

Spring Gentian *Gentiana verna*

The GENTIANACEAE, although found throughout the world are mainly temperate in distribution. In Ireland there are five genera containing eight species. The solitary, brilliant blue flower springs from a persistent rosette or cushion of generally oval but variable leaves. The total height of the plant varies only from 2 to 6cm of which the strongly-angled calyx takes up 1cm. The corolla is from 1.5 to 2.5cm in length and up to 2cm across. It is in flower from April to June and is pollinated by butterflies and moths.

Everyone's picture of a gentian is of a large, single flower of intense, heavenly blue, nestling close to the earth in an alpine environment. Many species of gentian conform to this picture but many do not. The Spring gentian, therefore, is of special importance in giving the conventional picture of gentian beauty.

Although absent from Wales and Scotland, Spring gentian is present in Yorkshire, Durham, Westmoreland and Cumberland, at elevations of 330 to 800m in limestone areas where it is stony and grassy. In Ireland in the counties of Clare, Mayo and Galway it is local but abundant where it grows, which is from sea-level to about 330m. Elsewhere it is found in the mountains of Central Europe, from the Sierra Nevada through the Caucasus to Mongolia. A sub-species occurs in Morocco.

Sea Bindweed
Calystegia soldanella

Irish CONVOLVULACEAE, of which Sea bindweed is an attractive member, consists of three genera — *Calystegia,* with three species including Common bindweed, *C.sepium*; *Convolvulus,* with one species, and *Cuscuta* also with one species, the parasitic Dodder, *C.epithymum. Convolvulus sepium* is a common, aggravating, garden weed. Worldwide, the CONVOLVULACEAE consists of about 1,000 species in about forty genera, and is predominantly

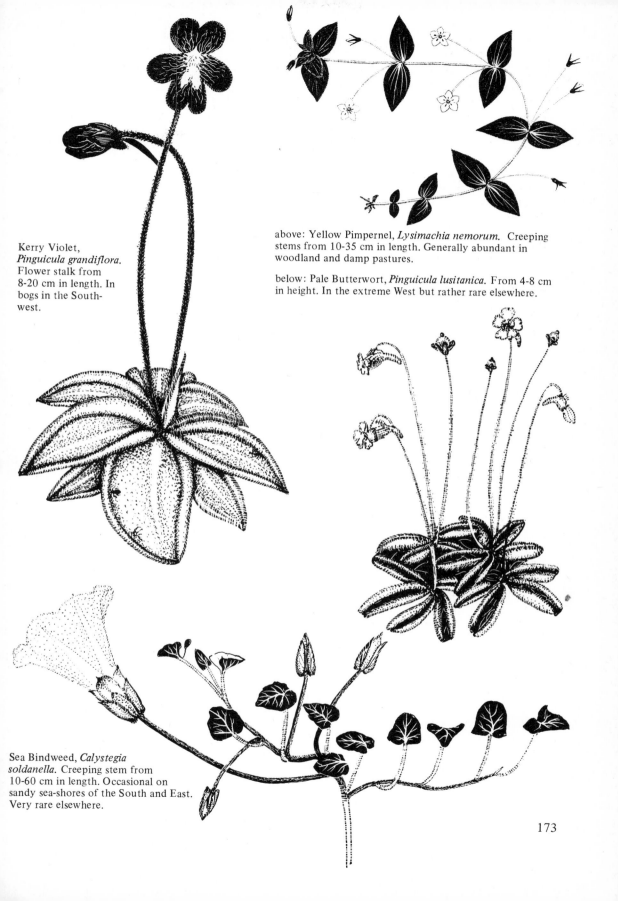

Kerry Violet,
Pinguicula grandiflora.
Flower stalk from
8-20 cm in length. In
bogs in the South-
west.

above: Yellow Pimpernel, *Lysimachia nemorum.* Creeping
stems from 10-35 cm in length. Generally abundant in
woodland and damp pastures.

below: Pale Butterwort, *Pinguicula lusitanica.* From 4-8 cm
in height. In the extreme West but rather rare elsewhere.

Sea Bindweed, *Calystegia
soldanella.* Creeping stem from
10-60 cm in length. Occasional on
sandy sea-shores of the South and East.
Very rare elsewhere.

tropical. These plants are characterized by their long, slender and often twining stems, their large flowers (cf. Dodder), and possession of two large leaflike bracts which enclose the calyx.

To the writer, however, Sea bindweed is an unfortunate name. It is true that the flower resembles that of a number of noxious weeds, to which it is closely related, but the large 32mm diameter, pale-pink, trumpet-shaped flowers seen amongst sand-dunes (accompanied by the much smaller, fleshy, kidney-shaped leaves), is a sight wholly beautiful. It is rightly appreciated by those who walk along shores and come across this plant when it blooms in June and July. Sea bindweed is an occasional inhabitant of the south and east coasts, only very rarely being found in the north and west of Ireland.

Dodder *Cuscuta epithymum*

This leafless, rootless annual is completely para-sitic on heather, *Calluna*, and gorse, *Ulex*. To the non-botanist it bears little resemblance to the CONVOLVULACEAE, to which it belongs in classification and which has already been described. The long, slender, red stems entwine amongst the host's shoots, adhering to them by suckers. The scented, pinkish flowers are about 2.5mm across and are borne in dense clusters from 5 to 10mm in diameter. Bellshaped and with four or five pointed lobes, the flower 'mouth' is partially closed by scale-like segments. It blooms from July to September. Dodder is very rare in Ireland, its localities being generally in southern coastal areas. The writer found the illustrated specimen on top of cliffs at Toe Head, Co. Cork.

Pale Butterwort *Pinguicula lusitanica*

The LENTIBULARIACEAE consists of five genera containing about 250 species world wide, and although they resemble members of the SCROPHULARIACEAE, they are easily dis-tinguished from the latter by their insectivorous habit.

In Ireland, two genera occur, *Pinguicula* with three species, and *Utricularia* of four species. The latter are known as the Bladder-worts, which are submerged aquatics with finely

Irish Spurge, *Euphorbia hyberna*. Grows to a height of about 60 cm. Almost entirely confined to West Cork and South Kerry.

divided leaves and small, bladder-like traps for capturing tiny, aquatic insects. No species is abundant. The three species of *Pinguicula* are known as butterworts and are of similar habit, in that the single flowers arise from a yellowish rosette of glandular leaves with in-curled margins. Pale butterwort is identified by its pale, yellow-ish or lilac-coloured, almost regular flowers and a short, blunt, downward-curved spur. It is only frequent in bogs and along stream-sides in the extreme west of Ireland, elsewhere it is rarely come across. Its distribution abroad includes West Scotland, South Wales, Isle of Man, South-west England, West France, West Spain, Portugal and North-west Morocco.

174

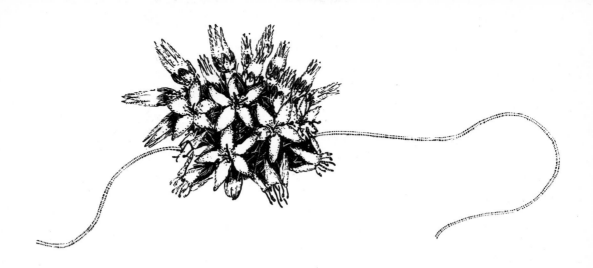

above: Dodder, *Cuscuta epithymum.* The dense clusters of flowers are from 5-10 mm across. Very rare, being found in a few localities in the South near the coast. Enlarged 5 times.

below (left): Lesser Skullcap, *Scutellaria minor.* About 15 cm in height. Found in the South, South-west and East.

below (right): Buck's Horn Plantain, *Plantago coronopus.* The rosette of leaves from 5-10 cm across. Perhaps our commonest sea-side plant.

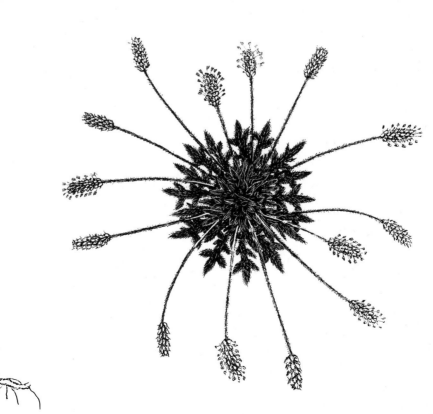

The Kerry Violet
Pinguicula grandiflora

Stated by Scully to be the most beautiful member of the Irish flora and by Praeger as being the most attractive of the Hiberno-Lusitanian species, this large butterwort grows in the greatest abundance (covering wet rocks, in bogs, damp pastures and roadsides) in Kerry and Cork. Northwards it becomes rare at the Shannon and, similarly, at Mallow to the east. In altitude it ascends from sea-level to 930m. Classified in the family LENTIBULARIACEAE, it is insectivorous as, indeed, are all the members of this family. Insects are caught by the glandular hairs, not only on the rosette of leaves, but also on the flower stem, the calyx and on the outside of the lower corolla lobe.

Outside Ireland it occurs in the Jura Mountains, the French Alps, the Pyrenees and at high altitudes in the north Spanish mountains. However, it naturalizes itself readily in suitable situations and is known in Cornwall, England, but it will sometimes disappear from a locality without obvious reasons. Hybrids occur between *P.grandiflora* and the Common butterwort, *P.vulgaris*, but as the latter is rare in *P.grandiflora* habitats, they are uncommon.

The plant is very like Common butterwort, *P.vulgaris*, but larger. The leaves, which are up to 8cm in length, are bright yellow-green in colour and form a tight rosette. The outer margin of the leaves curls inwards. The sepals of the calyx are joined, although the upper sepal is divided almost to the base. The violet corolla has a white patch in the throat and the shallow lobes of the lower lip are wider more so than longer, and the backwardly-directed (sometimes bifid) spur is at least 10mm in length. The flower-stalk may be from 8 to 20cm in length.

Lesser Skullcap *Scutellaria minor*

The LABIATAE is a fairly large and important family, members of which are usually easily recognized. The principal characters are the square-sectioned stem; the opposite, simple leaves without stipules; the inflorescences borne in the axil of opposite bracts (whorls); and a dense spike formed by the whorls at the apex of the stem. The corolla is basically tube-like but

with an upper lip composed of two lobes and a lower lip of three lobes — the central one being larger than the laterals. There are four stamens adhering to the corolla tube and two carpels containing two ovules each.

Worldwide there are about 3,000 species in

Creeping Willow, *Salix repens.* Seldom more than 1 m in height. Locally common in coastal areas. Rare in centre.

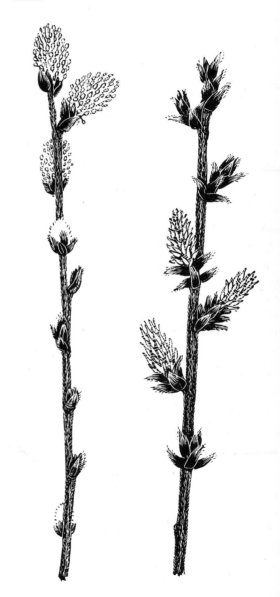

170 genera. The Irish Flora has nineteen genera containing thirty-nine species, many of which have an aromatic odour. The example illustrated is a small, perennial plant, scarcely 15cm in height with single, pinkish-purple flowers. The calyx is small and the corolla is about 8mm in length. The plant is in bloom in August and September. It occurs in marshes, bogs and wet heaths, but only in the south-west, south and east of the country — hardly ever being located elsewhere.

Buck's-horn Plantain
Plantago coronopus
The family PLANTAGINACEAE consists of only three genera and approximately 200 species, and although occurring throughout the world, it predominates in the north temperate region. In Ireland there are two genera, with six species in the genus *Plantago* and one in the aquatic genus *Litorella.* Two species, however, ribwort *P.lanceolata* and Great plantain, *P.major* are often pernicious weeds in cultivated or semi-cultivated places and would excite no-one. This is mainly because of the long, thick and strong tap-root which they possess, making them virtually impossible to uproot by hand. Hoary plantain, *P.media,* has been introduced into Ireland but still remains rare. Buck's-horn plantain, however, has attractions not possessed by others of this species.

The hairy leaves of this plant grow in a dense rosette, tight to the ground. They are pointed and bear a number of secondary, pointed lobes. The rosette may be from 5 to 10cm across and from the centre spring long-stalked, cylindrical blooms with individual flowers about 3mm across and brownish or yellowish-brown in colour, except for the slender-stalked, pale yellow stamens. It is entirely a coastal plant where almost every situation is colonized from rocks and crevices, between stones of harbour walls to sandy fields and roads. Buck's-horn plantain is perhaps our most common seaside plant.

Irish Spurge *Euphorbia hyberna*
The family EUPHORBIACEAE is a very remarkable one. It consists of over 280 genera containing about 7,000 species and is predominantly tropical in distribution. Many trees and shrubs are included, but all nine species found in Ireland, in two genera, are herbs. They all have simple leaves and the flowers are small, unisexual, lack a corolla — sometimes a calyx, and are generally green in colour. In the genus *Euphorbia,* with seven Irish species, a white juice (latex) flows from the injured plant. The leaves are alternate except for a whorl surrounding the inflorescence. The female flower is stalked and possesses three styles and is encompassed by a ring of single-stamened, male flowers. Each flower group is again encircled by minute, scale-like bracts and a ring of thickened, yellow glands.

Irish spurge is a slightly downy perennial, growing up to 60cm in height. There are five rays to the inflorescence, which is bright yellow in colour, enabling it to be viewed from a distance during April to June, when in bloom. It occurs in a variety of situations in West Cork and South Kerry.

Creeping Willow *Salix repens*
The family SALICACEAE consists of shrubs and trees of two genera, *Salix* — the willows and osiers, and *Populus,* the poplars and aspens. There are often difficulties in identification due to hybridization, especially in *Salix,* and due to introduction, as in *Populus.*

The Creeping willow is a creeping shrub, almost entirely coastal in distribution, and hardly ever attaining a height of more than a metre. Its woody stems, springing from a creeping rhizome, may stand erect or may be prostrate and are silky when young. The leaves are extremely variable in proportions and may be oval to long and spear-shaped and, again, when fresh are more or less covered with silky hairs, but only hairs on the undersurface are persistent. The leaves are from 12 to 36mm in length. The catkins are upright and either sessile or are borne on short stalks covered with small leaves, and they appear before the latter. The specimen illustrated was collected from Brittas Bay, Co. Wicklow, where large 'mats' of this species were growing amongst the sand dunes, and where it is browsed by the very large caterpillars of the Puss moth, *Dicranura vinula.*

16 Flowering Plants/Monocotyledons

The second group of flowering plants is the Monocotyledons. All the plants which have so far been described are Dicotyledons, and whilst the embryos of the latter possess two cotyledons or seed-leaves, the Monocotyledons have only one. There are several other important differences between these two main groups — such as the arrangement of internal tissues.

The leaves of Monocotyledons are usually parallel-veined and the floral parts are mostly in threes. The group contains 18 families found in Ireland and amongst them are found the highest development of both insect- and wind-pollinated plants. Additionally a substantial number are associated with water.

Ladies Tresses *Spiranthes spiralis*

Otherwise known as Autumn Ladies Tresses or Common Ladies Tresses this plant belongs to the family ORCHIDACEAE — one of the largest families of flowering plants. There are about 7,500 known species classified in about 450 genera. Although distributed throughout the world, it is predominantly a tropical group and their collection in tropical forests and their cultivation in hot-houses, has been the wealthy man's hobby. European species are, on the whole, much less 'showy' than those from the tropics. In Ireland, 28 species classified in 15 genera are to be found, although some of these are either rare or very local. Hybridization in some genera is common and there is a wide variation in others, so that there are often difficulties in identification.

The pattern of flower arrangement is characteristic. There are six perianth segments — the three outer (sepals) are usually similar, and the inner three (petals) are also similar except that the lower one, known as the labellum, is usually much broader than the other two. The single stamen is fused with the style, forming a central column, on the top of which is located the stigma. Behind the stigma or slightly above, lie two pollen masses known as the pollinia, and these arise often on slender stalks from the anther of the single stamen. The ovary lies beneath the point of insertion of the sepals and petals, and is often spirally twisted. The seeds are extremely small and numerous. Pollination occurs when the tongue of an insect seeking nectar withdraws with the pollinia adhering to it and on visiting another flower, the pollinia attach themselves to the stigma of the unfertilized bloom.

Autumn Ladies Tresses is a small, slender, inconspicuous plant with a rosette of leaves tight on the ground and with a flowering spike from 7.5 to 20cm in height. It is a native. The small, whitish flowers are arranged in a single spiral. The root consists of two or three tubers. It is in flower from August to September. It may be found, more or less infrequently, throughout the southern half of the country, but it occurs more widely amongst close turf near the edge of the sea in the south-western areas.

Early Purple Orchis *Orchis mascula*

This rather stout and solid-looking native plant — a member of the ORCHIDACEAE — grows to a height of from 12 to 35cm. The root consists of two globular tubers from which springs a rosette of oblong leaves which are often spotted. The flowers are reddish-purple and form a loose spike, although there is some variation in the degree of compactness of the blooms. The lateral sepals are upright, whilst the upper one is arched forward over the upper petals. The wide labellum is convex and three-lobed, with the central and larger lobe often notched. The backwards-directed spur is curved.

The Early Purple is in bloom from early April to late June, according to prevailing climate. Generally growing in damp pastureland, banks and road verges, there is a marked increase in its numbers after woodland has been cleared or coppiced. It prefers calcareous soils but is by no means confined to them.

Early Purple Orchis, *Orchis mascula*. Height from 12-35 cm. Damp pastures and road verges. Frequent.

Broad-leaved Marsh Orchid
Dactylorhiza majalis

Surely the most difficult genus of monocoty-ledonus plants – even, indeed, of *all* Irish flowering plants, for identification. One has only to look at the synonyms quoted in the modern *Orchid Flora* of D.M.T. Ettlinger (1976) or in Dr. Webb's *An Irish Flora* to realize this. But the plant referred to here is a marsh-land or water-meadow species and after exhaustive enquiries and examination of the specimen, the author confidently came up with the name of Southern marsh orchid, *O.praetermissa*, only to find that it had *not* been recorded from Ireland! He then had to fall back on the Broad-leaved marsh orchid, *D.majalis,* but the drawing was made from the originally collected plant, before identification. There is an Irish Broad-leaved marsh orchid, *D.m.occidentalis* described in the literature, and a synonym given as *D.kerryensis.*

The stem may be from 10 to 40cm in height and it is hollow to a limited extent. The leaves, which may be spotted, are dark-green and tapered at both ends, with the flowers of a light-mauve, to deep reddish-purple, and the bracts, especially of the lower blooms, noticeably leafy. The labellum is very variable in shape, although usually broad and flat, or even saucer-shaped. This plant is much more frequently seen in the west and north of the country than elsewhere, and it is in flower from mid-May to mid-June. This genus surely deserves a detailed examination by Irish botanists, and their findings should be published in a form which can be easily assimilated by naturalists everywhere.

Bee Orchid *Ophrys apifera*

It is hoped that the author will be forgiven for including both species of the genus *Ophrys* in this account of the Irish Flora, but each is so distinctive and unusual – although so closely related to each other and thrilling to find – that surely the genus deserves our special attention.

Bee orchid may send up from a pair of globular tubers, flowering, leafy spikes of about 40cm in height (but much less than this is more usual). There may be from three to six flowers in each spike, each widely separated. The sepals are pink, whilst the narrower and shorter upper petals are green or brown. The labellum, however, is distinctive – being very convex and velvety-brown, with yellow markings. The lobes of the labellum are not observed, as they are tucked underneath, which gives it a very bee-like appearance. The plant is in bloom during June and July, is relatively rare and occurs in dry, rough pasture, banks and sand dunes, where the soil is predominantly limestone.

Fly Orchid *Ophrys insectifera*

There is no mistaking this strange little inconspicuous member of the ORCHIDACEAE, which occurs almost exclusively on calcareous soils. It will grow up to 60cm in height in shady situations, but in more open areas it may be only 15cm. It is often found in marshy or boggy, scrubby places. The rootstock consists of two ovoid tubers and a few fleshy roots. Leaves arise from the flowering stem – but only a few flowers are borne on this. Three sepals are also present and they are of a bright-green colour. The reddish-brown upper petals are very narrow, whilst the three-lobed labellum is larger than the sepals with the centre lobe notched. Except for a blue patch in the centre, the labellum is also reddish-brown and looks much more like a small moth than a fly! It is rare in Ireland but does occur in the west and centre of the country. It is to be found in flower in June.

Yellow Flag *Iris pseudacorus*

He would be a very unobservant traveller in Ireland who was not aware of the patches of Yellow Flag in damp corners of meadows, ditch-sides, river-banks and marshes, during June and July. The long, sword-shaped, pale-green leaves may be up to 1m in length and the bright yellow flowers are usually about 8cm across with the outer perianth segments drooping. Two or three flowers are present on each stem but they open singly. The Yellow Flag is a member of the family IRIDACEAE and there are two other Irish species – Blue-eyed grass, *Sisyrinchium bermudiana,* and the Montbretia, *Tritonia crocosmiflora.* The latter is a garden hybrid which has become naturalized due to corms being thrown

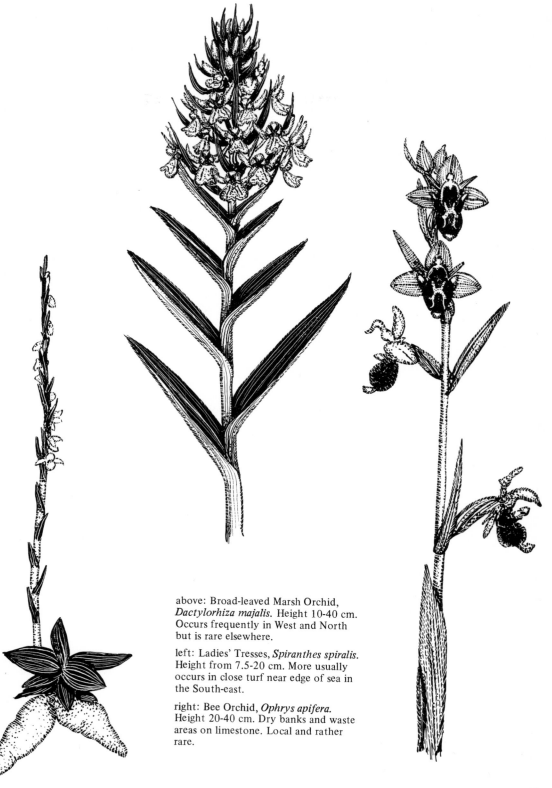

above: Broad-leaved Marsh Orchid, *Dactylorhiza majalis*. Height 10-40 cm. Occurs frequently in West and North but is rare elsewhere.

left: Ladies' Tresses, *Spiranthes spiralis*. Height from 7.5-20 cm. More usually occurs in close turf near edge of sea in the South-east.

right: Bee Orchid, *Ophrys apifera*. Height 20-40 cm. Dry banks and waste areas on limestone. Local and rather rare.

away with garden refuse. It is said also to spread by seed from time to time.

Kerry Lily *Simethis planifolia*

Ireland possesses six species of the LILIACEAE. Except for the bluebell, *Hyacinthoides non-scriptus,* and Bog asphodel, *Narthecium ossifragum,* they are rare or local. Perhaps the most interesting (and certainly the least known) is the Kerry lily, *Simethis planifolia.* It occurs only in a few, scattered places, all within about 50 sq.km around Derrynane in Co. Kerry. These precious plants are within a short distance from the sea where it is dry, heathy and rocky. Their blade-like, tapering leaves are up to about 25cm in length, although when the author saw the plant, after a cold, dry spring, they were only half this size. Only about twelve to twenty flowers are borne on each stalk, and the buds are purple, yet open to show a clear, white, star-like flower. This is truly one of Ireland's most precious plants.

Bog Asphodel *Narthecium ossifragum*

The lily family, or LILIACEAE, is distributed worldwide and consists of about 2,500 species contained in approximately 200 genera. In Ireland there are seven genera containing fourteen species.

Bog asphodel is a common inhabitant of boggy areas and is often abundant in wet mountainous and rocky places. The plant possesses rigid, grass-like but fleshy leaves arranged in two rows which arise from a slender (not bulbous) rootstock. The leaves are sometimes curved. The flowering stems grow up to 30cm in height and bear a few scale leaves, and a number of bright yellow flowers, each with six perianth segments. After flowering, the whole bloom and the stem turn a very attractive orange colour, and it is to be found in bloom from July to September.

Three-cornered Leek *or*
Triquetrous Garlic *Allium triquetrum*

Of the seven Irish genera of the LILIACEAE, six contain only one species. The genus *Allium* contains seven species — although four of these were introduced and have since become naturalized. Three-cornered leek is one of them. It is easy to identify at virtually any period of the year, on account of the sharply 3-angled stem. The leaves are long and grass-like — up to 10mm in width and bright green. From three to fifteen nodding flowers are carried in an umbel. They are bell-shaped and white, with a green line through the centre of each segment. Flowering during April through to June, it is of special interest as the seeds are dispersed by ants!

A native of the west Mediterranean region, it also occurs in southern Spain, Portugal and Morocco through to the western parts of Italy, Sicily and Tunisia. In Ireland it can be seen in the greatest profusion along the hedge-banks in south-west Cork, where the writer obtained the specimen illustrated, but although locally abundant, it occurs in a number of places elsewhere.

Lords and Ladies *or* Cuckoo Pint
Arum maculatum

The family ARACEAE consists of more than 1,000 species contained in about 115 genera. They are mostly distributed in the tropics. In some regions, species of ARACEAE predominate — such as along the muddy banks of the great rivers of tropical South America. Of the twenty species in Europe most are Mediterranean in distribution. Only two wild species are found in Ireland. These are the Lords and Ladies or Cuckoo Pint, *Arum maculatum,* and the Sweet Flag, *Acorus calamus,* and even the keen botanical observer could be forgiven for overlooking them. The arrangement of the inflorescence is quite exceptional and perhaps is best and most characteristically exemplified by Lords and Ladies.

In the first place, the most conspicuous and what appears to be part of the floral mechanism, is a large, pale-green, hood-like spathe. The lower part of the spathe is wrapped around the inflorescence proper. The latter is known as the spadix and consists of a column or upper part, usually dull-purple in colour and around the lower part there is a circle of hairs, then a zone of very small female flowers, the upper part of

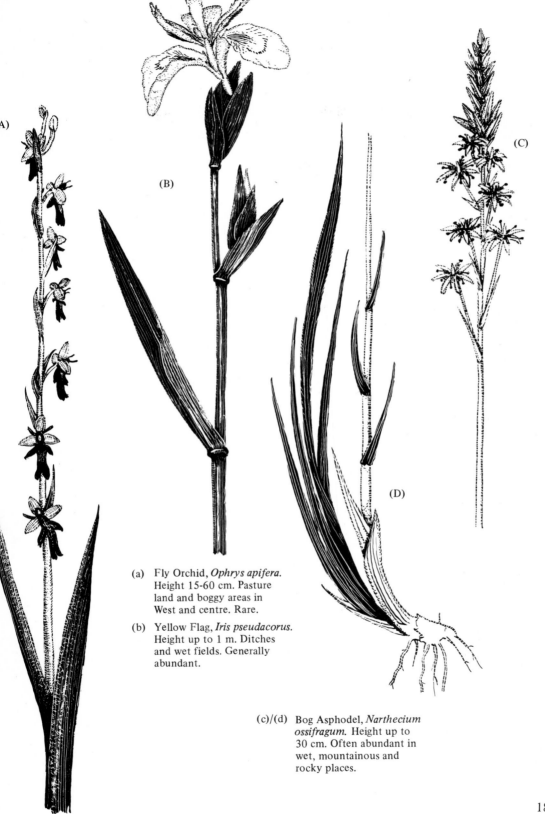

(A)

(B)

(C)

(D)

(a) Fly Orchid, *Ophrys apifera*.
     Height 15-60 cm. Pasture
     land and boggy areas in
     West and centre. Rare.

(b) Yellow Flag, *Iris pseudacorus*.
     Height up to 1 m. Ditches
     and wet fields. Generally
     abundant.

(c)/(d) Bog Asphodel, *Narthecium
         ossifragum.* Height up to
         30 cm. Often abundant in
         wet, mountainous and
         rocky places.

183

which contains sterile blooms. There is no perianth. Fertilization is carried out by small midges which become trapped in the lower part of the spathe and whilst wandering about over the lower parts of the spadix, transfer pollen from the male flowers to the female blooms. It flowers during April and May.

The fruits consist of a tightly-packed mass of bright scarlet berries which burst out from the light brown persistent base of the spathe. In spring the large, arrowhead-shaped leaves are conspicuous in the bottoms of shady hedges and in woodland. The leaves are net-veined – unusual in monocotyledonous plants, and are generally spotted with dark purple. It is widely distributed in Ireland, although found only in scattered localities. Growing in gardens and often in semi-wild conditions – especially in sheltered positions in the south-west – is the attractive 'Arum Lily', *Zantedeschia aethiopica,* with its pure white spathe and golden-yellow spadix. It is a native of South Africa.

Sea Arrow-Grass *Triglochin maritima*
Two species of Arrow-grass occur in Ireland. These are placed in the family JUNCAGINACEAE and are now the sole members of it – a third species, formerly only occurring in Co. Offaly, is no longer found there. Both species of Arrowgrass have grass-like, but fleshy leaves about 20cm in height. Stalkless flowers arise from a long, single stem, and the perianth segments are scale-like. Whereas *T.palustris* inhabits marshes and wet pastures, *T.maritima* is confined to salt-marshes and muddy shores where it is common. The flower of *T.palustris* bears six carpels whereas that of *T.maritime* bears only three.

Lesser Pond Sedge *Carex acutiformis*
The sedges, or family CYPERACEAE, are made up of about 300 species contained in approximately 100 genera, and they are worldwide in distribution. Ireland's sedge flora consists of eight genera and whilst in seven of the latter there are 23 species, in the single genus *Carex,* there are 67. They are generally said to be intermediate in habit and appearance – between

grasses, GRAMINEAE and Rushes, JUNCACEAE.

Sedges are usually indicative of land which is badly drained – marsh or generally wet areas, such as pond and ditch margins. They are characterized by their solid stems, which often are three-angled. The leaves, which may be alternately arranged or basal, sheath the stem at their lower ends. Unlike most grasses, this sheath is not split. Like grasses, however, the flowers are arranged in spikelets – either singly or in groups. Again, like the grasses, they are wind-pollinated – with a consequent reduction, or in some cases the complete absence of a perianth. Sedges, therefore, constitute a generally inconspicuous part of the scene, although viewed by the farmer with some misgivings.

Each spikelet is made up of a number of small, overlapping, scale-like bracts known as glumes. These are usually brown, and in the axils of some of them are single, stalkless, small flowers. In the genus *Carex,* the flowers are unisexual; in all other genera they are hermaphrodite. There are usually three stamens and although the style is single at its base, it divides into two or three stigmas. These are thread-like and slightly feathery.

Lesser pond sedge is fairly large, its creeping rhizome sending up stout, rough stems to a height of up to 120cm. It has flat leaves from 7 to 10mm in width. The top two (or sometimes more) dark-brown spikelets are male, whilst the light-brown, bottom two (or more) are female, although on occasion the central one may be of mixed sexes. The lower spikelet may have a long stalk. It is frequently found in the east, but is very rare in the west of the country and where it does occur it is found along ditches, riverbanks, lake-shores and in marshes.

Star Sedge *Carex echinata*
Star sedge mostly grows in dense tufts and the slender stems grow to a height of 35cm, although in exposed positions it may only reach 12cm. The long leaves are spiky and narrow, and the spikelets are about four in number and stalkless. The female flowers are up to eight in number and the male flowers occupy a position near the base. The rather flat-looking fruits eventually become star-like – pointing in all directions in

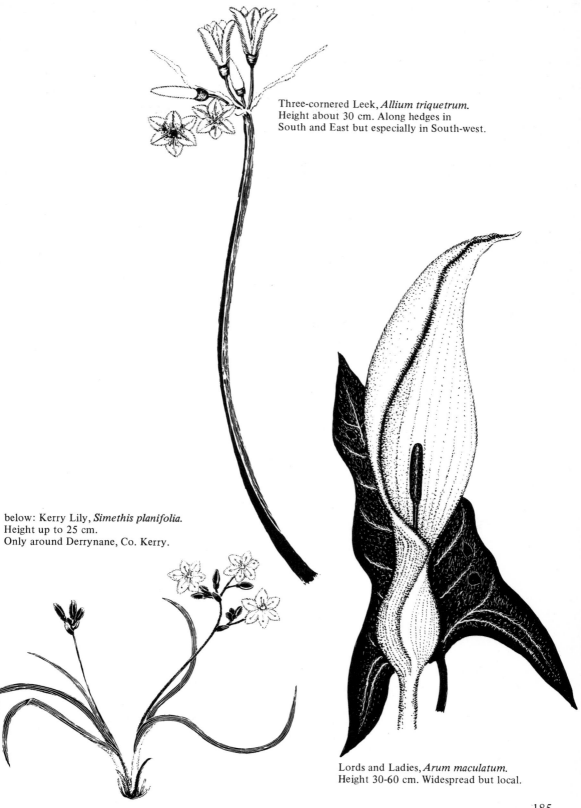

Three-cornered Leek, *Allium triquetrum*.
Height about 30 cm. Along hedges in
South and East but especially in South-west.

below: Kerry Lily, *Simethis planifolia*.
Height up to 25 cm.
Only around Derrynane, Co. Kerry.

Lords and Ladies, *Arum maculatum*.
Height 30-60 cm. Widespread but local.

which the glumes, in time, become inconspicuous. The fruit assumes a bifid tip and the edges of the beak are toothed.

This is a common Irish plant associated with bogs and moors, and yet may go unobserved by all except the specialist and the totally involved naturalist, but all peaty, mostly wet areas (from sea-level to almost the highest points) should be searched for it. The specific name is derived from the 'spikyness' of a sea-urchin.

Tufted Sedge *Scirpus caespitosus*
This common Irish sedge will often be encountered by the observant walker on boggy moorland and on mountain sides. It forms very dense and often large, dark-green tussocks and the stems are cylindrical, slender and un-branched. At the base there are a number of sheaths, which give the plant a dense, tufted habit. The pointed apex of the uppermost sheath, however, is elongated and green. There are about five flowers in the solitary, brown spikelet which is borne at the end of the stem, and the lowest glume is as long as the whole of the spikelet, terminating in a narrow green process.

Grasses *Gramineae*
The GRAMINEAE or Grasses constitute one of the most important groups of flowering plants, and is one of the largest families, consisting of about 600 genera containing approximately 10,000 species. They are found throughout the world — wherever flowering plants can live, and such significant cultivated species as wheat, *Triticum*; oats, *Avena*; rye, *Secale*; and maize, *Zea,* are all grasses. In addition, members of this group are of the utmost importance as forage for herbivorous, domestic animals — cattle and sheep. In Ireland, 43 genera are represented, in which there are 91 species, and a number of these are rare, and confined to very small areas.

Most grasses are perennial and are character-ized by their hollow, cylindrical (and usually simple) stems which are solid and swollen at the nodes. The narrow, long and pointed leaves are produced alternately in two series, and the lower part of each sheaths the stem. Generally, the sheath is split on the opposite side to the leaf's origin, and a small, membraneous appendage — the ligule — is present at the junction between the sheath and the leaf-blade.

Grass flowers are generally small and arranged in two series of from one to ten, mostly hermaphrodite, flowers. These are known as spikelets. The perianth is virtually absent consist-ing only of minute scales. Two bract-like *glumes* occur below the flowers or may enclose them. Grasses are invariably wind-pollinated, hence the absence of the brightly-coloured perianth and the possession of long filaments by the (usually three) stamens. Each flower of the spikelet is enclosed by two bracts, the inner of which is known as the *palea* and is usually smaller than the outer *lemma*. The superior ovary consists of a single cell and the two styles terminate in fairly conspicuous, long, feathery stigmas. The latter is characteristic of wind-pollinated flowers generally. The fruit of a grass, known as a caryopsis — in which the seed-coat is fused with the inner fruit layer — is usually enclosed in the *palea* and *lemma* which persist.

Marram Grass *Ammophila arenaria*
Marram grass is a large, robust and conspicuous plant. It is common around the whole coast of the country wherever there are sand-dunes. The creeping stems form an extensive network deep in the sand, rooting out at the nodes and often playing an important part in dune stabilization. On this account it has frequently been planted in coastal defence work. Growing to a height of 120cm, its leaves are rigid and terminate in a sharp point. The upper side of the leaf is green and shiny, but underneath it is white and ridged. During dry conditions, transpiration is restricted by the leaves rolling up, and thus moisture is conserved.

The whitish bloom may be up to 15cm in length and the spikelets densely arranged. The ligule is exceptionally long, with two pointed lobes. Around the north and south coasts (in a few areas), there could be some confusion in the identification of Marram, as the species Lyme-grass, *Leymus arenarius,* occurs on dunes and is superficially similar to Marram. However,

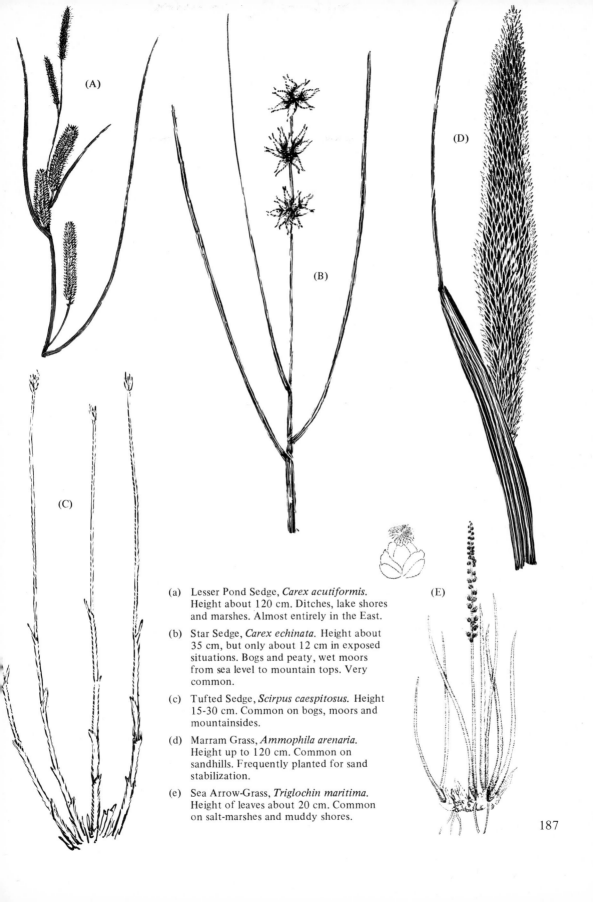

(A)

(B)

(C)

(D)

(E)

(a) Lesser Pond Sedge, *Carex acutiformis*. Height about 120 cm. Ditches, lake shores and marshes. Almost entirely in the East.

(b) Star Sedge, *Carex echinata*. Height about 35 cm, but only about 12 cm in exposed situations. Bogs and peaty, wet moors from sea level to mountain tops. Very common.

(c) Tufted Sedge, *Scirpus caespitosus*. Height 15-30 cm. Common on bogs, moors and mountainsides.

(d) Marram Grass, *Ammophila arenaria*. Height up to 120 cm. Common on sandhills. Frequently planted for sand stabilization.

(e) Sea Arrow-Grass, *Triglochin maritima*. Height of leaves about 20 cm. Common on salt-marshes and muddy shores.

187

the ligules in Lyme-grass contrast with those of the latter in being very short.

Vernal Grass *Anthoxanthum odoratum*
The rather tufted and hairy, slender stems of this grass grow to about 50cm in height and the leaves are flat and up to 5mm in width. The inflorescence is compact, shaggy and oblong, and the spikelets consist of only one fertile flower. The *glumes* are pointed but are not produced to the extent that they could be considered as awns. Between the *glumes* and the flower are additional hairy *glumes* which represent undeveloped flowers, and these do bear long awns inserted on their backs. The *lemma* and the *palea* do not, however, possess awns. Another common name for this grass is 'Sweet-scented' grass and this, with the specific name, refers to the pleasant, sweet odour of coumarin when it is drying. Vernal grass is plentiful wherever there are meadows, heaths and woods.

Yorkshire Fog *Holcus lanatus*
Yorkshire Fog is a down-covered grass, giving it a soft texture. It grows in tufts to about 60cm in height and usually forms a tussock. The pale green leaves are flat and up to 8mm in width and the lower sheaths are veined with reddish-purple and the ligule is only about 1mm in length. The spikelets are also downy and are about 4mm in length and maybe pink or pale green in colour. These form a feathery, pyramid-shaped panicle, up to about 13cm in length. Yorkshire Fog is an abundant grass of meadows and pasture-land generally as well as road margins and waste places. It is distributed throughout Europe and temperate Asia, and has been introduced into North America.

False Oat *Arrhenatherum elatius*
The False oat grows to about 120cm in height from a creeping rootstock. Two forms of the plant occur; in one there are bulb-like swellings at the base of the stem; in the other form these swellings are absent. At one time it was thought that two distinct species were involved but it is now known that they hybridize freely and a range of intermediates occurs.

The form, in which the swellings are present, is known as *tuberosum* or sometimes *var bulbosus*. The specimens from West Cork measure about 10mm in diameter. The leaves are few in number – in width, up to 8mm and rough to touch. The inflorescence may be as long as 25cm and often droops. The spikelets consist of an upper hermaphrodite flower and a lower male one. The West Cork specimens were blotched with purple but no reference to this has been found in literature. The lower *lemma* bears a long awn on its back, which is bent. The upper *lemma* is provided with a short awn only or this may be absent altogether. The West Cork specimens had no awn. False Oat appears to favour banks and hedges and various types of waste ground. At Dromreagh, Co. Cork, it covers an old, stone bank hedge where a cluster of 'bulbs' can always be found.

Dog's Tail *or* Thraneen Grass *Cynosurus cristatus*
A wiry, erect grass, it may be found growing to a height of about 60cm with 2 to 3mm wide, flat leaves. It is easily identified by the inflorescence, which is a tapered spike – up to about 8cm in length. The spikelets are densely arranged on one side only of the stem, which is sinuate particularly at the base. They are also flattened and in small groups – the basal ones being sterile and reduced to a feather-like row of *glumes*. The fertile spikelets consist of from three to six flowers, and the *lemmas* are furnished with a long point, but, being somewhat less than 1mm in length, cannot be considered an awn. This is an abundant grass, widely distributed in grassland and roadside verges, and growing in a wide variety of soils.

Wall Barley *Hordeum murinum*
This annual grass, which grows to a height of from 15 to 50cm, has rather a strange distribution. It occurs in central and southern Europe, North Africa and western Asia. In addition, it is frequently found in North America. In the British Isles it is common in southern England,

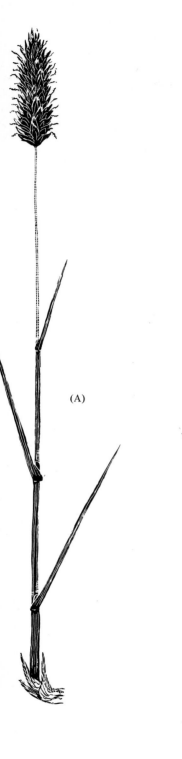

(a) Vernal Grass, *Anthoxanthum odoratum.* Height up to about 50 cm. Plentiful in meadows, heaths and woods.

(b) Yorkshire Fog, *Holcus lanatus.* Height up to about 60 cm. An abundant grass in pastures and associated areas.

(c) False Oat, *Arrhenatherum elatius.* Height up to about 120 cm. Banks, hedges and wasteground often in abundance.

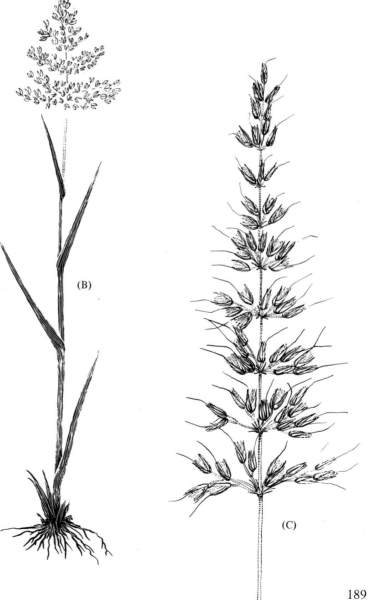

(A)

(B)

(C)

but it extends into Scotland — although being rare throughout much of the north. In Ireland it appears to have arrived recently. It can only be regarded as a coastal species — very rare, except along the east coast, and yet it is abundant around Dublin! The specimen illustrated was growing alongside a tarmacadam path near the centre of this city.

The blades of the short, rather hairy, rough, short leaves are no more than 5mm in width. The flower spike may be anything from 4 to 10cm in length, tapering slightly and consists of spikelets in groups of three flowers. In each spikelet, which is up to 12mm in length, there are both sterile and hermaphrodite components. The rough awns are long and conspicuous. Wall barley is not a grass of benefit to agriculture, but is most often found as a rather harsh grass at the base of walls and in rocky waste places, as indicated both by its common and specific names. Almost certainly it has been introduced into Ireland accidentally, through agriculture. Commuters to Dublin should have no difficulty in finding this grassy alien.

Sand Couch *Elymus farctus*
formerly *Agropyron junceiforme*
Sand Couch grows to about 50cm in height from mat-like rhizomes which grow on new sand-dunes. The genus *Elymus* includes the notorious Scutch or Couch-grass, *E.repens,* a persistent weed of gardens, very difficult to eradicate except by chemical means.

The flowering stems are stiff and erect and the plant is greyish-blue in colour, with leaves which are finely ridged and roll inwards when dry, being more or less flat when damp. This mechanism for slowing down or restricting transpiration is commonly found amongst grasses growing in dry situations such as sand. In Sand Couch, the leaf sheath is persistent and creamy white. This plant is very frequently found on sandy sea-shores around the whole of Ireland's coastline.

False Oat 'Bulbs', *Arrhenatherum elatius.*

190

Dog's Tail, *Cynosurus cristatus*. Height up to about 60 cm. Widely distributed and abundant in grassland and roadside verges.

Sand Couch, *Elymus farctus*. Height to about 50 cm. Often found on new sand-dunes around the entire coast-line.

Wall Barley, *Hordeum murinum*. Height from 15-50 cm. A rare, coastal species abundant in Dublin. A recent, accidental introduction.

17 Conifers

CONIFERS *Gymnospermae*

The Gymnosperns are naked-seeded i.e. the seeds are not completely surrounded by the ovary. They are mostly evergreen trees and shrubs. The native flora of Ireland has three species only: Scots pine, *Pinus sylvestris*; juniper, *Juniperus communis*; and yew, *Taxus baccata*. None of the species plays any significant part in the total indigenous Irish flora. However, nine genera and about 180 species make up the family PINACEAE, and many are of great importance in their native lands where softwood timber is converted from them. In Ireland some of the species have been extensively planted by the Forestry Department — most important being Japanese larch, *Larix leptolepsis*; Sitka spruce, *Picea sitchensis*; Corsican pine, *Pinus nigra* var. *laricio*; and Lodgepole pine, *Pinus contorta*. A number of other species are also planted in estates and in parks, and have become well-known.

Douglas Fir *Pseudotsuga taxifolia*

The 'Douglas' fir is not a native Irish tree but grows naturally on the Pacific coastal belt of North America — from British Columbia to California. Its usefulness as a plantation, forest-grown tree has been outstanding. Today, however, it has been surpassed by other species, yet it accounted for about a half million young trees planted in Ireland annually during the first half of this decade.

It has an interesting history. It was first collected as a specimen by Archibald Menzies in 1792 when sailing with Captain Vancouver, but it was David Douglas who first sent seeds to Britain in 1825. These were obtained from Fort Vancouver on the Columbia River.

The tree grows to an immense size — some reaching 100m and with a 13m girth. The forester expects it to grow about 60cm annually for fifty years. The tree, when growing in the open, is a beautifully proportioned pyramid, with the lowest branches bent to the ground under their weight. Higher up, the branches are horizontal and feathery, but the topmost branches are ascending. The bark is smooth and dark-grey in colour, with resin blisters, but in old trees the bark becomes very thick and rough. The long, pointed, reddish-brown buds bear a remarkable resemblance to those of beech, *Fagus sylvatica*. The 2.5cm long needles are parallel-sided, with a rounded apex, and they arise spirally from the twig, although during growth they twist around to form a flat spray. The flowers are inconspicuous, but the female grows to a cone, tapering towards base and apex, which may be up to 10cm in length and hangs downwards. It is perhaps most characteristic of all cones, which makes identification a very simple task. This is on account of the three-tongued bract with which each scale is provided. The three tongues are pointed and the central one is longest and thickest. (The cone of Douglas fir is a 'Fir' cone but many people refer to pine cones as fir cones.)

It has been the much-valued experience of the author to have seen Douglas fir planted, and then, fifty years later, to have seen them felled at 30m for timber.

Sitka Spruce *Picea sitchensis*

The two most widely planted conifers in Ireland are Sitka spruce and Lodgepole pine, *Pinus contorta.* Over the last few years, about 64% of all plantings has been of the former species and about 22% of the latter. Sitka Spruce is much like Norway spruce, *Picea abies* — well-known as the Christmas tree, but it is easily distinguished from it. Sitka is a native of British Columbia and Alaska, and is considered to be a valuable forest tree, with a wider range of adaptation to growing conditions and a greater production timberwise than Norway spruce.

In its native land, Sitka grows to about 70m, but in Ireland it generally attains only half of this, with a diameter of trunk from 1 to 2m. The inner branches and twigs are not shed when dead. The thin bark is grey or purplish-grey and divides into rounded scales which have a tendency to peel off. The leaves are spirally arranged, flattened and stiff — although they tend to stand out all around the stem, the upper leaves point slightly forwards and hide the stem. The long, thin leaves are from 12 to 18mm in length with a sharp point at the apex. The upper surface is a bright, yellowish-green, whilst the lower is silvery. The leaf is also 'keeled' — making it triangular in cross section. The cone is cylindrical and pale-brown in colour, and is only from 5 to 10cm in length compared with 10 to 15cm in the case of Norway spruce. *Pinus contorta* is distinguished from Scots pine by the needles being short — maximum length 6cm — whereas in the latter the maximum needle length is 10cm, and the needles are thin and twisted.

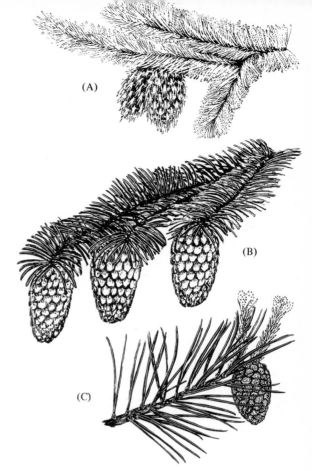

(a) Douglas Fir, *Pseudotsuga taxifolia.* Although growing to a height of 100 m on the Pacific coast of North America, plantation trees in Ireland are expected to reach 30 m in 50 years growth.

(b) Sitka Spruce, *Picea sitchensis.* In Ireland plantation trees attain a height of about 35 m. A native of British Columbia and Alaska. The most widely planted tree in Ireland.

(c) Scots Pine, *Pinus sylvestris.* Reaches a height of about 30 m. Not thought to be indigenous but is frequently found naturalized in the West.

(d) Yew, *Taxus baccata.* Sometimes attains 20 m in height. Rare as a native tree, probably due to attempts to eradicate on account of its poisonous nature to cattle. Occasionally bird-sown.

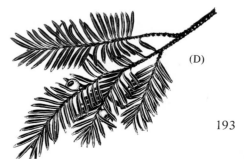

193

Japanese Larch *Larix leptolepsis*

Since about 1910 a species of larch, *Larix leptolepsis,* which grows wild on the mountains of Japan, has attracted much attention as a forestry tree because its growth rate is faster and its adaptability to climatic conditions better than many other plantation species. As a consequence, a high proportion of larch plantings by our Forestry Department are of this species. However, the species readily cross-fertilize, and the first cross progeny show fast and strong growth. This "hybrid vigour" as it is called, is a common phenomenon. The hybrid is known as *Larix eurolepsis* and is commonly planted.

The European larch, *Larix europaea,* is not a native Irish tree but originates in the central European mountain ranges. It has, however, been planted extensively in the western countries of Europe for 300 years or so. It is a common Irish tree in parks, gardens and forestry plantations.

Larches are deciduous, the pale, straw-coloured needles falling in autumn (except the terminal needles on the twigs of very young trees). The young needles are put forth in spring in tufts at the tip of knob-like, short shoots. Those of Japanese larch are bluish-green, whilst those of the European larch are bright, emerald green, gradually dulling as the season progresses. Japanese larch also has rusty-red twigs; European larch has straw-coloured ones. Male flowers are produced just as the needles emerge and consist of clusters of yellow anthers. The female flowers are attractive, flower-like cones of yellow or pink colour. The cones of the two species allow easy identification. In the case of Japanese larch, the cone scales are reflexed, but straight in the European species. The timber of larch has superior strength properties to those of most commonly-grown conifers and the heart wood, which is reddish-brown in colour, is naturally durable. Larch species account for between two and three per cent of all trees planted in Ireland.

Scots Pine *Pinus sylvestris*

Scots pine, otherwise known in Ireland as Scotch fir, is one of the most widespread tree species in the north temperate regions of the Old World. In Europe, it extends from Norway and Sweden in the west, to the Kamchatka Peninsula on the Pacific Coast of Russia facing the Bering Sea in the east. Southwards it occurs in Southern Spain, Northern Italy and Albania. In Ireland, however, it is not thought to be indigenous, although it has been extensively planted and is naturalized in a number of areas — more frequently in the west of the country.

Scots pine grows to a tall tree and may reach 30m in height. The bark is characteristic. The upper part of the trunk is a bright, reddish-brown, nearly orange, in fact, whilst the lower part is dark brown. The upper bark is shed in thin, scale-like plates, whilst the lower bark is longitudinally fissured, forming irregular plates. There is a pair of blue-green, long (3 to 8cm) spine-like, twisted leaves arising from each 'short shoot'. The cones are from 3 to 7cm in length and dull-brown in colour. The seeds possess a membranous wing which is about three times its length. The male cones are yellow and occur in clusters around the base of the new year's growth, whilst the female cones are

Juniper, *Juniperus communis ssp nana.* Prostrate, falling over rocks. About 1 m in height. Local in the West and North on acid-type rocks.

Japanese Larch, *Larix leptolepsis*. Reaches a height of about 30 m. Planted extensively by the Forestry Dept.

reddish and terminal. The latter consist of numerous scales, each bearing two seeds. In the second year the cones are green and in the third, brown and woody.

Juniper *Juniperus communis*

A bushy, evergreen shrub with prickly, needle-shaped leaves it is perhaps the best known for producing the black berries from which the flavouring for gin is obtained. Actually, the berry-like structures are cones, each containing three seeds. The scales in the young female cone swell to form fleshy 'berries' which are at first green in colour, later turning to black. The leaves may be up to 2.5cm in length but are often mucn smaller. There are two sub-species.

In *communis* the leaves, which are narrow and sharp, spread so that the form of the shrub may be erect or prostrate — this sub-species is generally found on limestone. In *nana*, which has at one time been given specific rank as *J.sibirica*, the form is *always* prostrate — with the leaves incurved — thus giving the young shoots the appearance of cones. The leaves of *nana* are not so sharp as in *communis* and the shrub is found only on acid-type rocks.

Yew *Taxus baccata*

The yew is a widely-known tree, not only throughout Europe but also as far east as the Western Himalayas and as far south as the North African mountain ranges. When fully-grown it has a rounded outline and a height of some

20m. The trunk is massive, often fluted, and with leafy twigs springing from it — even to the base. The thin bark is reddish-brown and scaly. The leaves are from 1 to 3cm in length, with small stalks, and are dark green on the upper surface and much lighter below. The margins curve downwards. The leaves are borne spirally, but they twist around, forming two opposite rows.

The unisexual flowers are almost entirely on separate trees — the male being a globular group of stamens, and produced in the leaf axels of the twigs from the previous summer. The female is a naked ovule surrounded by small bracts. After fertilization, a fleshy red or yellow 'cup' grows from the disc on which the ovule develops — almost enveloping it — and this is known as the aril. It flowers during March and April.

Irish yew, *T.fastigiata*, is a well-known variety of the Common yew, and grows with an upright habit — the branches and twigs also being erect. The leaves do not twist into two opposite rows along the twig. Two female trees were originally found in the mountains of Fermanagh around 1780. One planted in the garden of the discoverer, a Mr. Willis, died in 1865 but the other tree, planted in Florence Court, Fermanagh, is the progenitor of all Irish yews.

In Ireland, the yew is a relatively rare tree. Usually it is found only in inaccessible places, such as on cliff-faces and on islands, otherwise it ornaments private gardens or public parks. It is most frequent in the West — extending eastwards to Cork — but it also occurs in Co. Antrim and Co. Longford, and although found in Co. Wicklow, this may have originated by planting.

195

18 Ferns

FERNS *Pteridophyta*

In many parts of Ireland the ferns and their related groups are a conspicuous element of the flora, and whilst damp and shady localities are usually associated with their growth, it must be remembered that even in relatively dry situations (such as neglected pastures), the Bracken fern is often the dominant species of vegetation. In addition Rustyback and Wall-rue can grow in almost desert conditions on mortared walls. The general observer or naturalist often shies away from identification, and whilst it is true that a number of species require a close and detailed examination in order to differentiate them, at least 15 species are so distinct that their naming should provide no difficulties to the general reader, and many of these are illustrated.

Ferns, together with mosses, fungi, lichens and algae, are *all* flowerless and form the group known as the Cryptogams. The flowering plants are termed the Phanerogams. Whereas the latter produce seeds, the Cryptogams multiply by the much smaller bodies known as spores. Ferns differ also from the other Cryptogamic groups by their possession of specialized cells arranged in strands, which not only give some degree of rigidity to the various structures, but allow sap to be conducted through the tissues. The ferns and their allies are, for this reason, known as the Vascular Cryptogams.

The leaf-like organs spring from a stem which, although showing much variation, is generally a somewhat woody rhizome or rootstock. The stem is often encrusted with the bases of old fronds. The leaf-like organs, known as fronds, may be simple or relatively so, or subdivided (compound). In the latter, the central stalk continuation bears a number of pinnae on each side and in turn these may be similarly divided, the resulting segments being called 'pinnules'. The pinnules, in some cases, are divided also. The ferns are unlike the flowering plants in the carrying of the reproduction organs on the fronds. The spore-producing organs are known as sporangia, and are produced in clusters called 'sori' (singular — sorus). Often the sorus is protected by a membrane called an indusium, the shape of which is important in identification.

The life-cycle of ferns is of great interest. When a spore is shed, it germinates if it falls onto a suitable moist substratum. Often some degree of shelter must be present. It then grows into a flat, scale-like plant known as a 'prothallus', and this is usually only about 1cm across. It is long-living if suitable conditions prevail. Sexual organs are produced on the underside of the prothallus and spermatozoids are released from the male organ, one of which fuses with the egg cell of the female organ. From the fused cell a fern develops. Such a life-cycle involving a sexual stage produced on the prothallus and an asexual stage where spores are produced, is known as 'alternation of generations'. This is of great importance to those who wish to have an understanding of the relationship which the various groups of living organisms bear to each other.

Ferns and their allies — *Pteridophyta*

Family		Number of species
Ophioglossaceae	Adder's Tongue	2
Osmundaceae	Royal ferns	1
Hymenophyllaceae	Filmy ferns	3
Salvinaceae	Azolla	1
Marsileaceae	Pilularia	1
Polypodiaceae	Polypody, Bracken, Buckler ferns, Male ferns, Shield fern, Hard fern, Spleenworts, Hart's tongue, Rustyback, etc.	31
Equisetaceae	Horsetails	8
Lycopodiaceae	Club-moss	3
Selaginellaceae	Fir-moss	1
Isoetaceae	Isoetes	2

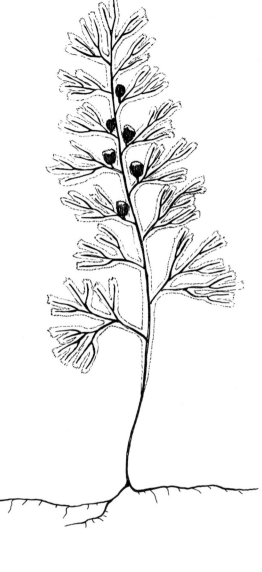

above: Tunbridge Filmy Fern, *Hymenophyllum tunbridgense*. Fronds 8 cm in length. Locally in damp, sheltered places mainly in the lowlands.

left: Royal Fern, *Osmunda regalis*. Nearly 2 m in height. Grows widely in the West, less common in the East.

197

Royal Fern *Osmunda regalis*

The sole representative of its family in Ireland, this fern is usually outstanding in the habitats where it occurs. The stout, erect fronds are nearly 1.5m in height, and whilst the outer fronds are entirely green and leaf-like, the inner ones differ in having the upper pinnules consisting only of masses of sporangia, brown in colour. The Royal fern occurs widely in the West, in wet situations such as in ditches around the edges of bogs, on river banks, as well as in damp woodland. It is not so common in the East.

The Filmy Ferns *Hymenophyllum spp.*

These very small, moss-like ferns are found only in damp situations and also where there is shelter. The fronds, which are not more than about 8cm in length, are only one cell in thickness and are semi-transparent. Two species are found in Ireland, The Tunbridge Filmy fern, *Hymenophyllum tunbrigense*, and Wilson's Filmy fern, *Hymenophyllum wilsonii.*

The Polypody Family

Of all the ferns, 85% belong to this family, and although most would be called typical ferns with pinnately divided fronds, some possess fronds of a much more simple pattern. There are 13 Irish genera, which can be identified by the shape and position of the sori or by the presence or absence of an indusium; and the species can often be distinguished by the characters of the sterile frond.

Bracken *Pteridium aquilinium*

This is one of the best known of all ferns and is worldwide in distribution, being found from the Arctic to the tropics. It occurs abundantly throughout Ireland in many habitat types. It does not, however, flourish in limestone areas, and when it occurs it is usually stunted in growth.

The fronds, which may be as much as 2m in length if supported by hedgerows, are produced from a very long, underground rhizome. The curled-up fronds force their way through the soil in spring and by the end of June to July are fully expanded. They change colour in

autumn, and during winter fall to the ground. Although the form and colour of bracken is often appreciated by those walking in the countryside, it is a serious nuisance to the farmer as it diminishes, to a considerable extent, the grazing potentiality of the land, and it is difficult to eradicate.

Maidenhair Fern, *Adiantum capillus-veneris*. Fronds up to about 30 cm in length. Mainly in the Burren of Co. Clare.

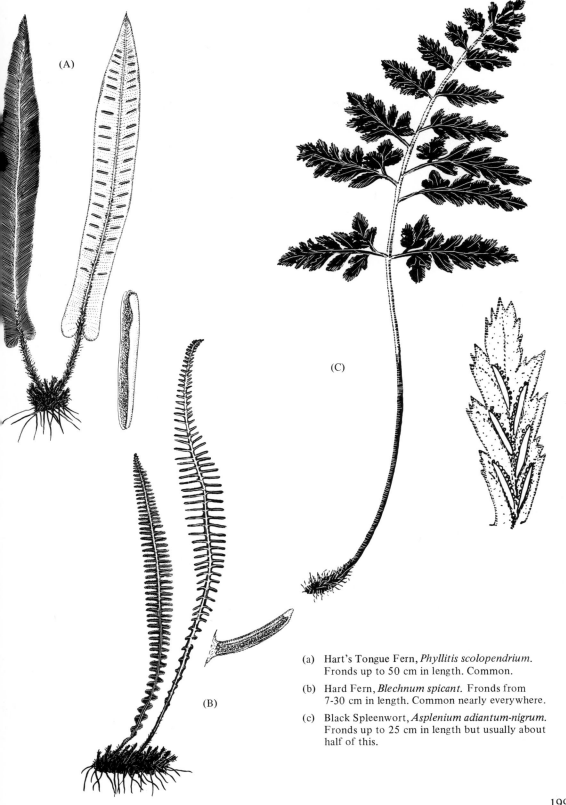

(A)

(C)

(B)

(a) Hart's Tongue Fern, *Phyllitis scolopendrium.*
Fronds up to 50 cm in length. Common.

(b) Hard Fern, *Blechnum spicant.* Fronds from
7-30 cm in length. Common nearly everywhere.

(c) Black Spleenwort, *Asplenium adiantum-nigrum.*
Fronds up to 25 cm in length but usually about
half of this.

Maidenhair Fern
Adiantum capillus-veneris
This delicate fern must be familiar to many; not, alas, from a knowledge of its native haunts but from cultivation as a pot plant. The pale green frond segments are borne on long, shining black stalks which are from 8 to 30cm in length, whilst the creeping rhizome is covered with dark brown scales. The individual pinnules are fan-shaped, with the curved, outer margin cleft, the sides straight and all borne on slender, black stalks. In the fertile pinnules the sori are produced on the inner side of the margins where they are protected by flaps folded over from the pinnules. The Maidenhair fern is to be found in crevices and damp faces of rocks, from Clare to Donegal. For the most part it is rare, but in the Burren and in the Aran Islands it is less so.

The Hard Fern *Blechnum spicant*
This easily identifiable fern is common throughout Ireland wherever its preferred habitat occurs. This is woodland, hedge-banks, moorland and mountainous regions where there are acid conditions and some degree of shade. There are two distinct types of frond. The sterile frond is long, narrow, tapering gradually from its short, glossy-brown stalk, appearing in the spring and then persisting nearly throughout the following winter. The blade is divided into two rows of simple pinnae, like a comb.

The fertile frond is erect and longer than the sterile frond, appearing later in the year. The average frond length is 15 to 25cm. Its pinnae are fewer in number and much narrower, carrying a single sorus on each side of the midrib of each pinna. The sorus is almost as long as the pinna and covered by an inwards-opening, white indusium. When grasped, the Hard fern is harsh and rough to the touch, which gives it its popular name.

Hart's Tongue Fern
Phyllitis scolopendrium
This fern is easily identified by its long, undivided, strap-shaped fronds. These are variable in length growing from 15cm to about 50cm and up to about 8cm in width. The stalk is clothed with narrow, pointed brown scales, and at the base of the blade there are two ear-like basal lobes. The veins are parallel and on the underside of the blade the long sori (containing the sporangia) lie in pairs parallel to the veins. Although showing a preference for limestone, it is not confined to such areas, and is one of the most abundant ferns in hedgebanks.

Black Spleenwort
Asplenium adiantum-nigrum
There are about 700 species of *Asplenium* and they have a world-wide distribution. In Ireland there are eight species to be found. Spleenworts have a tufted rootstock with the bases of the previous year's fronds persistent and covered with narrow, brown, pointed scales. The stalk of Black spleenwort is long — half the total length of the frond — being dark-brown, glossy and black at the base. The frond varies from 2cm to 25cm in length and the blade is bipinnate and widest at the base. The pinnae have a triangular outline and are acute at the tip. Black spleenwort is found growing from crevices in stone walls and in shady, stone-faced hedgebanks. Some of the rarer species of spleenwort are confined to the south or west.

Maidenhair Spleenwort
Asplenium trichomanes
The rootstock of this fern is similar to that of the Black spleenwort, but when old may form a dense, compact tuft. The fronds are from 5 to 20cm or so but the stalk is short, brown or almost black, wiry and glossy, with the pinnate blade up to about 2cm in width, tapering slightly towards the base. The numerous pinnae are blunt, oblong and attached to the stem by their midribs only. On the underside of each, are from four to eight sori, located on the veins and each covered by an indusium. When the frond dies, only the pinnae drops off the stalk with the frond midrib persisting. It occurs in much the same places as Black spleenwort, and is very common especially in shaded areas and sheltered situations.

Rusty-back Fern *Ceterach officinarum*
This easily-identifiable little gem of a fern is widely distributed but is confined, however, only to limestone areas in Ireland where it grows in crevices in the rocks, but is much better known as growing on stone and brick walls where lime-mortar has been used! This, of course, very often occurs at great distances from the limestone areas already mentioned. The fronds, which arise as a dense tuft are from 3 to 20cm in length, and sometimes larger. They are characteristically shaped as a series of simple triangular segments along the main stem. The undersurface of the fronds is covered with pointed scales which overlap. They are silver-coloured at first but become a rich chestnut colour later. There is a network of veins in the frond segments, which is unusual.

Polypody *Polypodium vulgare*
Polypody fern occurs throughout the temperate regions of the world in both north and south hemispheres. In Ireland it is generally common, but found more widely in the west wherever shady, damp conditions prevail. Although growing on hedgebanks, in rock crevices and on old walls, it is perhaps better known as an inhabitant of tree-forks and of mossy, tree branches and trunks. In the latter situation, in damp woods, the rhizomes often form a dense, tangled mat over the bark.

It cannot be mistaken for any other Irish fern. The fronds are of variable size — from 5 to as much as 750cm in length, arising from the rhizome in two rows, at intervals, and the yellowish, scale-less stalk occupies from one-fifth to one-half of the frond length. The blade is usually lanceolate consisting of simple, rather leathery segments which are arranged along the midrib alternately. The sori containing the sporangia are circular, about 2.5mm in diameter and arranged in rows on the underside of the segments. Most often the sori occur only on the upper half of the blade but sometimes they are to be found on all the frond segments. The young sori are yellow, but later (in the autumn) they turn a bright orange colour, and after the spores are shed, the frond remains green until the following spring or summer.

Botanists now recognise three species of Polypody as occurring in Ireland. Southern polypody, *P.australe,* occurs in South Europe and in Ireland is found mainly on limestone outcrops and walls. It has broadly triangular fronds. Common polypody, *P.vulgare,* itself has much narrower linear fronds. It is a species of northern Europe and in Ireland is confined to acid rocks, walls and trees. *P.interjectum* has ovate fronds. It is the most widespread of the three species in Ireland. It is found both on acid and basic rocks and on trees and is the most common species on roadside walls and bridges.

Moonwort *Botrychium lunaria*
The frond of this fern is up to about 8cm in length and is divided into about six pairs of fan-shaped pinnae. The fertile spike is generally longer and irregularly branched. Although widely distributed throughout Ireland, it is generally rare, being found in old grassland in hollows and in sand dunes, and also on mountain ledges where, as in the case of Adder's Tongue, cultivation has not penetrated.

Adder's Tongue Fern
Ophioglossum vulgatum
This small fern is very un-fernlike in appearance, its oval frond being quite undivided and with the fertile, double-rowed sporangia-bearing spike standing up in front of it. It is characteristic in appearance although it is often difficult to find in old pastures and meadowland and rocky places where it occurs. It has recently become a marker to show where land has never known the plough, and sheep are said to relish it.

Dwarf Adder's Tongue
Ophioglossum lusitanicum
This fern resembles the Adder's Tongue in miniature, but there are some important distinguishing characters. There is a single, strap-shaped sterile frond scarcely ever more than 3cm in length and 5mm in width. There are no free-ending short veins as are present in the Common Adder's Tongue. The fertile spike bears only about six sporangia on each side. The root-

stock is tuberous. One important growth character is that the fronds appear in autumn and spores are produced during the winter, so that the plants are dying away just when Common Adder's Tongue is appearing. It is predominantly a Mediterranean species, but is recorded from the Scilly and Channel Islands as well as from Ireland. Amongst short turf on sea-cliffs is said to be the situation where it is most likely to be found.

(a) Dwarf Adder's Tongue Fern, *Ophioglossum lusitanicum.* Frond seldom longer than 3 cm. Short turf on sea cliffs in the South.

(b) Rusty-back Fern, *Ceterach officinarum.* Fronds from 3-20 cm in length. Mostly known from growing in lime-mortar in brick and stone walls.

(c) Polypody, *Polypodium vulgare.* Fronds from 5 to as much as 750 cm in length in very favourable conditions. Generally common where damp conditions prevail although more abundant in the West.

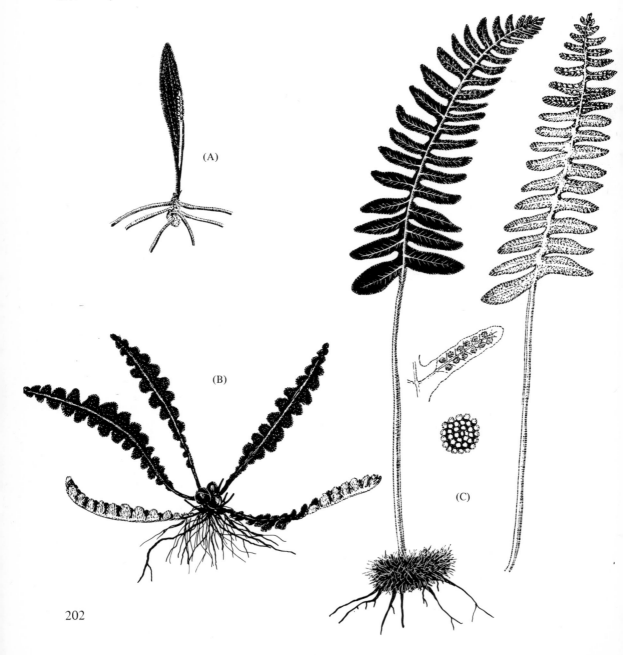

(A)

(B)

(C)

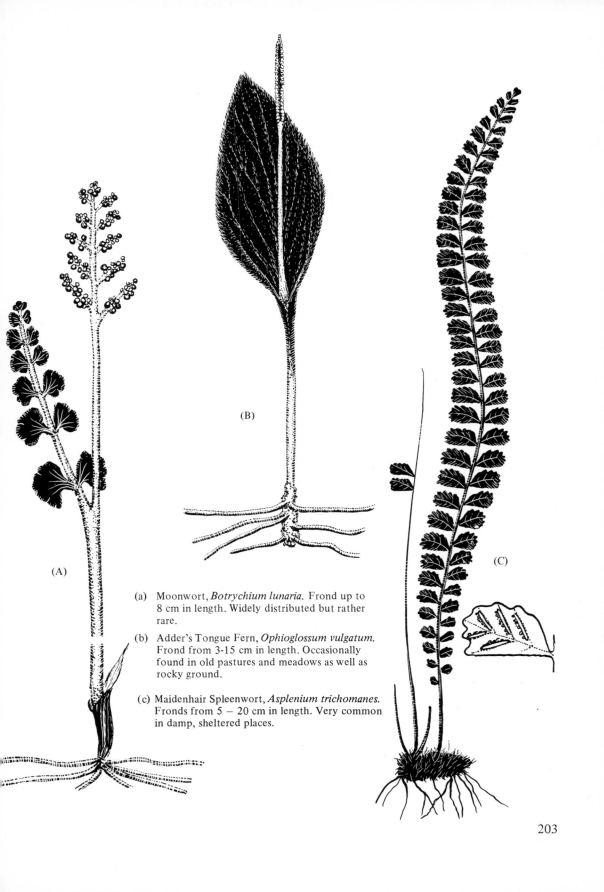

(A)

(B)

(C)

(a) Moonwort, *Botrychium lunaria.* Frond up to 8 cm in length. Widely distributed but rather rare.

(b) Adder's Tongue Fern, *Ophioglossum vulgatum.* Frond from 3-15 cm in length. Occasionally found in old pastures and meadows as well as rocky ground.

(c) Maidenhair Spleenwort, *Asplenium trichomanes.* Fronds from 5 – 20 cm in length. Very common in damp, sheltered places.

19 Fungi

Mushrooms, Toadstools, Bracket Fungi

The fungi are in a very special group of plants. They differ from all other plants in that they do not possess the green pigment chlorophyll, and therefore are not able to carry out the photosynthetic process which enables them to convert the atmospheric gas carbon dioxide and water, with the aid of energy derived from sunlight, into more complex carbohydrates such as sugars, starches and cellulose. This is with the exception that a limited number of species of flowering plants have degenerated into the saprophytic or parasitic habit, or rely on an intimate association with fungi and, therefore, do not possess chlorophyll and are unable to build up their own carbohydrates unaided. Some fungi are greenish in colour, but this is due to pigments other than chlorophyll.

Most fungi obtain their nutrients directly from dead organic material and are known as saprophytes. A number of these decay wood, whereas others feed on substances of animal origin. Others, again, are parasitic, obtaining their sustenance from living plants and animals. The fungi, therefore, constitute an essential element in nature. They also differ fundamentally from other plant groups in respect of their structure; their processes of reproduction, as well as their mode of nutrition.

The vegetative parts of the fungus plant consists of simple, thread-like rows of cells. Each filament is known as a hypha (plural – hyphae), and when they proliferate into a sheet or mass, it is known as the mycellium. The cell walls are seldom of cellulose but chemically are similar to the substance chitin, which makes up the external skeleton of insects. The vegetative phase of the fungus is seldom much in evidence, but the fruiting body, known as the sporophore (or spore-bearing organ) is often conspicuous. At least this is the case in the group known as *Basidiomycetes*, which includes most of the larger Fungi. When speaking of mushrooms, toadstools, bracket fungi and the like, we are referring to the spore-bearing body of the fungus – the vegetative threads largely go unnoticed.

Fungi are classified according to the manner in which the spores are produced. Besides the *Basidiomycetes*, the other main group is known as the *Ascomycetes*. In the latter, the spores are produced in a vase-like structure, the ascus, and there are usually eight spores in each, although four are sometimes produced. They are distributed by means of an explosive mechanism during abrupt changes of humidity. The *Basidiomycetes*, on the other hand, produce spores in fours, each of which is borne on a stalk outside a special club-shaped cell. The spore-bearing surface may be situated in gills, often sheet-like, beneath a cap or pileus – as in the mushroom, or may be located in tubes, as in *Boletus*. There are other variants too, and almost everyone is familiar with the 'puff-balls' in which the spores are expelled by the rupturing of the outer skin.

In such a highly specialized group as the fungi, there are a number of intricate anatomical variations, all with special names. In such a short survey as this present one, it is not proposed to include these but to describe and illustrate a number of examples of the better known Irish species. Fungi have always excited interest in connection with their culinary attributes and there are a number of species renowned throughout Europe for their flavour and palatability. Others, however, are known for their disagreeable effects, even lethal consequences, if consumed. In the notes on the examples which follow, this aspect of their association with man is given.

above: Leopard Amanita, *Amanita pantherina*. Cap 5-10 cm in diameter. Often found in woodland in addition to pastures and heaths. Highly poisonous.

left: Parasol Mushroom, *Lepiota procera*. Up to about 25 cm in height and sometimes attains this in diameter of the cap. Common around edges of woodland, especially in the vicinity of dead tree roots. Edible.

below: *Hygrophorus conicus*. Cap is about 2.5 cm across. Probably very common and widespread. Unwise to use in kitchen.

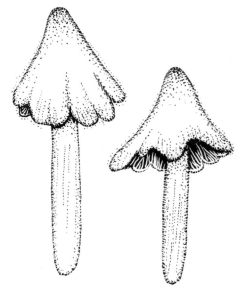

Leopard Amanita *Amanita pantherina*

This is a highly poisonous species, and in France where a much greater selection of fungal species is eaten than is in Ireland, it is on the 'dangerous' list. It is as well for the gatherer of wild mushrooms to remember its description. The cap is usually from 5 to 10cm in diameter and is olive-brown or smoky-brown and is round at first but flattens as it enlarges. The chief characteristic, however, is the white, persistent wart-like fragmentation with which the top surface of the cap is ornamented. The stem narrows towards the top and a ring encircles it near the centre, whilst just above the swollen base of the stem are several concentric rings. The flesh is white and remains so even when damaged, which contrasts with the related and similar species, *A.rubescens,* which becomes reddish when injured. The Leopard Amanita may occur in a variety of habitats in woodland situations, but may also be found in pastureland and on heaths.

Parasol Mushroom *Lepiota procera*

The genus *Lepiota* contains many species, but some of them are so characteristic that they can be recognized with ease. The most well-known is the large and handsome Parasol mushroom. The mushroom grows to about 25cm in height. The cap may be 10 to 20cm in diameter, and is greyish or creamy-brown and covered with rough scales. When it first bursts through the ground it is globular, but becomes bell-shaped, finally flattening to a parasol shape with a prominent knob. When young, the cap is covered with a velvety-brown felt, but as it enlarges the latter breaks up into small pieces, giving a scaly appearance, but the central knob remains brown. The fringed margin turns downwards. The base of the stem is usually rather swollen and there is a white, movable ring around the stem beneath the cap. The white flesh is thin and soft, and the Parasol mushroom is one of the most important of the edible Fungi.

It occurs most commonly around the edges of woodland, but when found elsewhere the indication is of old tree-roots where clearings have been made. The Shaggy Parasol, *Lepiota rhacodes,* is similar to the above species but is somewhat smaller, sturdier and with the scales

(A)

(a) Common Ink-cap, *Coprinus atramentarius.* May attain about 20 cm in height. Generally in and about woodlands and gardens. Edible but alcohol not to be taken at same meal.

(b) Puff Ball, *Lycoperdon perlatum.* Usually from 2-8 cm in diameter. Common in a variety of situations. Edible when young.

(c) Chanterelle, *Cantharellus cibarius.* Cap from 1-8 cm in diameter. Common under Beech and Oak trees. In the first class of edible fungi.

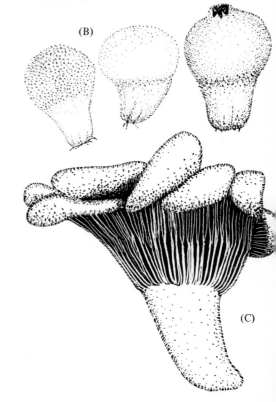

(B)

(C)

larger and rougher. This occurs in woodland and also in gardens, especially in rich ground on old compost heaps. *Lepiota acutesquamosa* bears sharp-pointed warts on the cap, which fall off later, leaving a series of scars. Of the three species given above, the Parasol mushroom is the most palatable.

Hygrophorus conicus

This brilliantly coloured fungus is small, of a bright, shining yellow with an orange peak to the sharply conical cap. It is very slimy, making handling difficult. The cap may be up to about 5cm across and the margin splits and may curve upwards. The gills are a pale yellow. The straight stem is marked longitudinally with small fibre elements. The whole fungus assumes a black colour when damaged or with age. The specimen illustrated was collected from upland, heathery moor at about 200m in September. It is inadvisable to use in the kitchen.

Lactarius blennius

Members of the genus *Lactarius* are chiefly characterized by the milky juice (latex) which they exude when damaged. The latex is produced by cap and stem. The cap of *L.blennius* is up to 10cm across and is slimy and fleshy, pale olive to greenish-grey in colour, with concentric rows of darker marks. When mature the cap has a central depression. The stem is pale grey, about 5cm in length. Its gills are thin and white. Most *Lactarius* species are very acrid to taste.

The Horse Mushroom *Psalliota arvensis*

This is a large species with the cap up to 20cm across. It is almost globular at first and creamy-white in colour but then it becomes flattened. It is soft and smooth to the touch, but handling gives rise to yellow and brownish bruises. The stem is up to 12cm in height and 2.5cm in diameter, and is swollen at the base. The gills are white at first but gradually turn to dark chocolate-brown. The flesh is white and firm and does not darken and the strong 'mushroom' smell is unmistakable. The Horse mushroom is often found in pastureland, especially near cow-sheds and hayricks. It has a rather stronger flavour than the Common mushroom but is nevertheless widely eaten.

Common Ink-Cap *Coprinus atramentarius*

Ink-caps are easily identified by their habit of auto-digestion. The whole cap of the fungus dissolves into a black liquid which consists of spores in a watery medium. The process starts at the outer edge of the cap, continuing inwards and often takes only a few hours to complete. The Common Ink-cap is soft-grey in colour, bell-shaped with white stalks. It usually occurs in small clusters. Both the Common Ink-cap and the Shaggy Ink-cap, *Coprinus comatus*, appear to live on decaying, buried tree-branches and are most often to be found on made-up ground, such as new foot-paths and lawns. The Shaggy Ink-cap may be found in very large numbers, yet growing singly, in such situations. Ink caps are edible with a distinctive, if somewhat, strong flavour. It is recommended that alcohol is not taken at the same meal.

Chanterelle *Cantharellus cibarius*

This completely egg-yellow, fungus is often common under beech and oak trees in autumn. It is fairly easy to identify on account of its colour, shape and apricot smell when fresh. When mature, the cap is depressed in the centre, exposing the broad, shallow and blunt gills. The stem narrows towards the base. The Chanterelle is an important component of Continental cooking, but for those wishing to use it in the kitchen, it should be borne in mind that it requires long, slow cooking — unlike the Common mushroom. The only species with which it can be confused is *Hygrophoropsis aurantiaca,* which has thin, unfolded gills and no odour, and is not to be recommended as part of one's diet.

The Penny Bun *Boletus edulis*

Fungi in the genus *Boletus* are for the most part large and fleshy, with a stout, central stem but gills are absent. The place of the latter is taken by a mass of downwards-directed tubes in which the spores are produced. If the cap is broken across, the tube gives the appearance of wood-grain. In this species the cap is up to 15cm across, hemispherical and tawny to yellowish-brown in colour, and with a granular finish, rather like the crust of a loaf of bread. The stem is thick — up to about 6cm in diameter and 12cm in height. Sometimes it may be pale brown with whitish areas, but covered with a fine network of raised, white lines, as though lace had been stretched over it, making it a very beautiful sight. The tubes under the cap are, at first, white but become greenish-yellow later. This species is edible as are other Boleti in browns and yellows, but the more colourful species should be avoided.

Dryad's Saddle *Polyporus squamosus*
This conspicuous bracket fungus is large, usually very large — up to 35cm across. It grows in a vertical series of brackets, each extending above the other from the trunks of living trees or the apparently dead stumps of broadleaved species. Elm is most commonly attacked. In shape the bracket is depressed in the centre and is semi-circular or fan-like. It is yellow in colour and patterned with brown. Feather-like scales, brown in colour, are arranged concentrically. The stem is short and thick but may take a variety of forms, but is black below and pale above. Although polypore fungi are usually persistent and with age become extremely hard, the Dryad's Saddle is an exception. At the end of autumn it dries up into dark-brown, torn strips and is then blown away by the weather.

Coriolus versicolor
This surely must be the most common of all the Bracket Fungi. These latter are the species which grow out horizontally from a tree tunk or branch and are often plate-like in form. Practically all the species of the POLYPORACEAE, as they are termed, derive their subsistence from the constituents of wood — cellulose especially. It is this, in the main, that brings about the decay of timber. *Coriolus versicolor* occurs widely on a variety of tree species, being found also on stumps, fence posts and fallen branches.

It consists of a semi-circular, thin plate marked with a beautiful concentric pattern of wavy lines in many colours, ranging from grey, greenish, yellow, brown and to black. The underside is white and covered with minute pores. It is not usually more than 8cm across, but frequently large numbers appear together and overlap each other, and all have a leathery consistency.

Puff Ball *Lycoperdon perlatum*
The Common Puff-ball is almost universally known — both in the creamy-coloured, young state and later, when the brown, paper-like skin bursts at the top to release the smoke-like spores. In shape, it is like an inverted pear and the top is covered with soft, spine-like processes. These, however, soon fall off but leave scars, each of which is surrounded by a ring of warts. When the flesh is still white and firm the puff-ball is edible, as is the Giant Puff-ball, *Calvatia gigantea*, which usually reaches football size.

Warty Earth-Ball *Scleroderma verrucosus*
Although placed last amongst the fungi, our two species of Earth-ball exhibit no special mechanisms for spore dispersal. Reliance is placed on their being set free by weather disruption or breaking open irregularly. It is about 5cm in diameter but somewhat flattened on top (in the specimen illustrated two were found growing closely together). The base is rather firmly attached to the mycelium and, if the Earth-ball is removed carefully, the white, sometimes thick mycelium, may be observed. At first the outer skin is covered with brown scales but these drop off at maturity. When cut through, it is found to be deep brown in colour inside — very different from a Puff-ball — and is inedible.

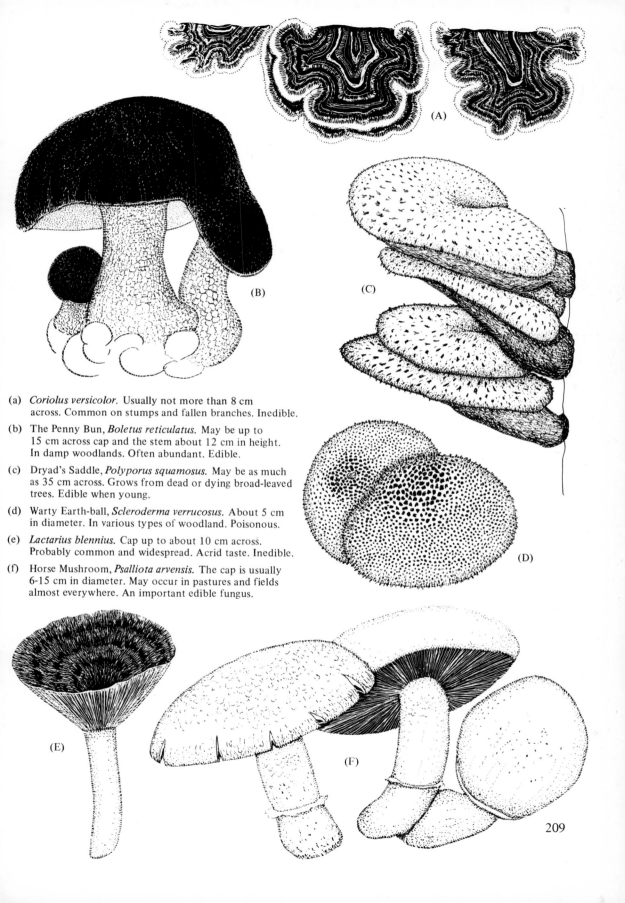

(A)

(B)

(C)

(D)

(a) *Coriolus versicolor.* Usually not more than 8 cm across. Common on stumps and fallen branches. Inedible.

(b) The Penny Bun, *Boletus reticulatus.* May be up to 15 cm across cap and the stem about 12 cm in height. In damp woodlands. Often abundant. Edible.

(c) Dryad's Saddle, *Polyporus squamosus.* May be as much as 35 cm across. Grows from dead or dying broad-leaved trees. Edible when young.

(d) Warty Earth-ball, *Scleroderma verrucosus.* About 5 cm in diameter. In various types of woodland. Poisonous.

(e) *Lactarius blennius.* Cap up to about 10 cm across. Probably common and widespread. Acrid taste. Inedible.

(f) Horse Mushroom, *Psalliota arvensis.* The cap is usually 6-15 cm in diameter. May occur in pastures and fields almost everywhere. An important edible fungus.

(E)

(F)

20 Mosses

Most mosses are small and some are very small, so that structural details are difficult to observe without x10 lens. The parts of a moss are as follows:

There is a stem which is usually slender and delicate and not more than 10cm long, though it is occasionally strong and wiry and up to 1m. It may be simple or branched, erect or prostrate. Fine filamentous rhizoids, which look like miniature roots, arise from the stem and these often serve to anchor the plant to the soil, rock or bank on which it is growing. Occasionally the rhizoids form a dense felt entirely covering the stem. Numerous simple, stalkless leaves arise from the stem and these are neither divided nor lobed. Except for a central nerve they are only one cell thick.

The life history of mosses resembles that of the ferns in the visible alternation of generations, sexual and asexual. The minute sexual reproductive organs are borne on the leafy plant, surrounded by specialized leaves at the apex of the stem or on short lateral shoots. The 'male' reproductive organs, the antheridia, which produce the spermatozoids, are cylindrical. The 'female' reproductive organs, the archegonia, are flask-shaped and each contains a single egg cell, accessible by a neck canal. Antheridia and archegonia may occur on the same plant (synoecious or autoecious mosses) or on separate plants (dioecious mosses).

The asexual generation consists of a capsule borne on a stalk or seta. It develops from the fertilised egg cell and remains attached to, and partly parasitic on, the sexual generation for the whole of its existence. Capsules are usually egg-shaped or cylindrical. They are green when young and are covered during development by a calyptra, formed from the upper part of the archegonium. As the capsule ripens it turns brown and the calyptra falls off, revealing the lid. The cells of the capsule wall dry out and the lid separates from the capsule wall, often by means of an annulus, a ring of cells which peels away like the strip of cellophane on a cigarette packet. The mouth of the capsule is usually guarded by the peristome made up of a ring of tooth-like segments. In wet weather, when spores would stick together and fall directly to the ground, the peristome remains closed. In dry weather, when the spores can be carried on convection currents, the teeth curl outwards, opening the mouth of the capsule. Small local changes in the humidity of the air cause jerky movements of the peristome teeth, which flick the spores away from the capsule. After being dispersed, in dry conditions, the spore germinates in damp weather, first forming a filamentous growth (the protonema), which then develops into the moss plant in the form in which it is generally known.

Polytrichum juniperinum

All mosses of the genus *Polytrichum* have a characteristic appearance. Their leaves are rather sharply divided into two parts — a broad, basal, whitish section which sheaths the stem, and a spear-shaped, outer section which is usually toothed. In *Polytrichum juniperinum,* however, only the very tip of the leaf is toothed and a lens is required to see this. The margins of the leaf roll inwards in dry conditions, giving the stem a spiky appearance. The tip of the leaf is reddish and the seta is bright red, and from 2 to 5cm long. The calyptra is very hairy (this is the derivation of the generic name) and it completely covers the capsule.

The capsule is box-like, distinctly quadrangular with a ring-like neck and crimson lid with a beak. Spore dispersal in the genus is also characteristic. When the lid is shed a white membrane remains. There are 32 or 64 stome teeth but their tips remain united with the membrane. The spores are shed through the gaps separating the central parts of the teeth. This species commonly occurs on acid heathland, especially in drier situations and is a fairly regular component of the plant succession after heathland has been burned.

Dicranella heteromalla

This common moss is generally found in areas of acidic soil, but has a wide variety of habitats — from heathland and woodland, to mountain slopes. The non-expert will usually be able to identify it from its pale, greenish-yellow setae supporting the orange capsules. The sickle-shaped leaves taper to a fine point, and all point in the same direction. The vein is broad at the base but tapers, ending some distance from the leaf tip which shows delicate teeth. The seta is from 10 to 20mm in length and the pear-shaped capsule is almost horizontal when mature. The lid bears a beak-like prolongation and the peristome teeth are red.

Dicranum majus

This moss is characteristic of woodland composed of Sessile oak, *Quercus petraea,* especially in rocky glens. It can be distinguished from most other mosses when in fruit by the setae which arise from the stem in bunches of up to six. The leaves are long and pointed — 10 to 14mm in length, sickle-shaped and generally all looping in the same direction. The long, fine tip of the leaf, which the vein does not reach, is delicately toothed, but a lens is required to see this. The seta is straw-coloured and the curved, dark-greenish-brown capsule bears a number of slight furrows. The specimen illustrated was collected in Glengarriff woods, Co. Cork.

Leucobryum glaucum

This is a moss which can be recognised from a distance by its colour and growing habit. It can only be mistaken for one other species, *L.juniperoidum.* It forms dense cushions, which may be up to 1m across, although the height of the plants individually is not much more than 10cm. The base of the cushion consists of partially decayed, lower parts of the shoots, which are white but the tip of each is a pale, greyish or whitish green. The specific name *glaucum* is a reference to this colour. If the cushion becomes dry the colour increases in whiteness. It is usually a bog-plant.

The leaves are variable in length but reach about 1cm. The nerve takes up nearly the whole width of the broad base of the leaf. Its small capsules are only rarely found as it appears to reproduce mostly by the terminal shoots producing a mass of purplish rhizoids which break away to form a new cushion.

Bryum argenteum

This again is an easy moss to identify and the town-dweller does not have to travel far to find it! It is very common in pavement cracks and the joins between slabs, being abundant even when the atmospheric pollution is so heavy that other bryophytes are precluded. The small cushions or cushion 'strips' are grey-green and are formed of tightly-packed, upright stems. Small, concave leaves overlap and are pressed to the stem, giving a catkin-like appearance. The leaves have a broad point of insertion in the stem and rarely reach more than 1.5mm in length, the

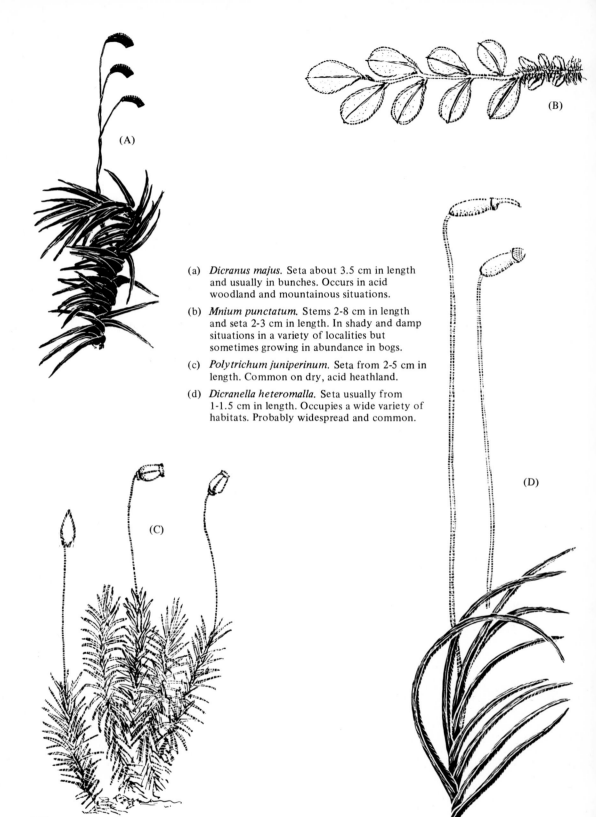

(A)

(B)

(a) *Dicranus majus.* Seta about 3.5 cm in length and usually in bunches. Occurs in acid woodland and mountainous situations.

(b) *Mnium punctatum.* Stems 2-8 cm in length and seta 2-3 cm in length. In shady and damp situations in a variety of localities but sometimes growing in abundance in bogs.

(c) *Polytrichum juniperinum.* Seta from 2-5 cm in length. Common on dry, acid heathland.

(d) *Dicranella heteromalla.* Seta usually from 1-1.5 cm in length. Occupies a wide variety of habitats. Probably widespread and common.

(D)

(C)

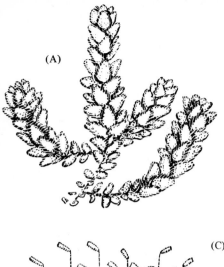

(A)

(C)

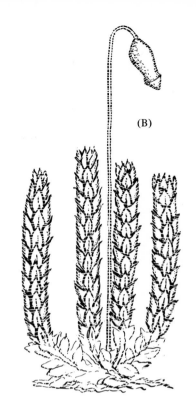

(B)

(a) *Hookeria lucens.* Seta about 2 cm in length. Only found in situations of shade, shelter and high humidity and grows on soil. Specimen illustrated was collected in Glengariff, Co. Cork.

(b) *Bryum argenteum.* Seta only about 1 cm in length. Mostly found around paving stones, roofs and walls in towns. Abundant everywhere where such situations exist.

(c) *Ptychomitrium polyphyllum.* Seta 7-15 mm in length. Colonizer of hard, siliceous rocks and also found on wall tops. Plentiful.

(d) *Leucobryum glaucum.* Height of single plant from 3 to about 10 cm but 'cushion' may be up to 1 m across. On acid, woodland floor or wet moorland. Also found on trees in districts of heavy rainfall.

(e) *Ulota crispa.* Seta only 1-2 mm in length. Occurs on branches of trees in high rainfall areas.

(E)

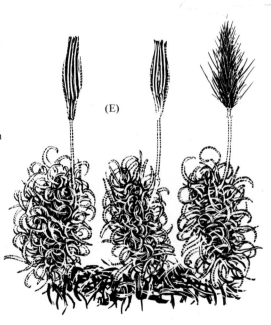

(D)

vein stopping short of the leaf tip which is pointed and prolonged. The capsule is small, only 1.5mm in length and of a reddish-brown to a deep crimson colour. The lid is vermilion and the capsule is borne on a short seta. This moss is usually found in man-made habitats but is also known from around gull colonies, and it is thought to prefer nitrogen-rich situations.

Mnium punctatum

This is a most distinctive moss which should not be difficult to identify. It is upright and the leaves are large — up to 8mm long and 5mm wide, being narrow at the base, broadest at about halfway and blunt at the tip except for a very small point at the apex (the apiculus). The broad nerve sometimes does not quite reach the apex. The leaf has a border of thickened cells which undulate slightly but teeth are absent.

The stem reaches about 8cm, the lower part being covered with reddish-brown rhizoids known as tomentum. The largest leaves form a loose rosette at the apex. The leaves are small at the base when very young and the shoots are light green, but with age they darken and may become reddish. The capsule is carried on an orange seta and the lid has a long beak. This moss is found in shady, wet situations such as the stony areas bordering streams, rock clefts, decaying tree stumps in damp woodland and sometimes in bogs where it flourishes.

Ptychomitrium polyphyllum

This is a colonising moss of hard, siliceous rocks where it forms dense, hemispherical cushions — tops of walls may also be colonised. It is a most abundant moss in many mountainous regions. The upper part of each stem is dull green, the lower part black. The pale, yellowish-brown capsules are a conspicuous feature. The cushions are generally about 3cm high but the leaves are long — from 3 to 5mm. *Ptychomitrium polyphyllum* can be identified with certainty by three microscopic features. Firstly, the leaf is longitudinally wrinkled or folded, so that furrows are observed on the surface. Secondly, the leaf-tip is bluntly notched. Thirdly, the cells near the leaf-base are characteristic, having thick, longitudinal walls, whilst the transverse walls are narrow. The capsules are said to be upright but in the specimen drawn they were slightly inclined. The fine peristome teeth of the capsule are red. The specimen was a dried plant from Glengarriff, Co. Cork.

Ulota crispa

Members of this genus usually grow on trees and *U.crispa* is common on tree-branches in mountainous areas. The small, round tufts or cushions are often abundant. The stems are less than 1cm long and the leaves of the upper part are light yellowish-green, but below are blackish-brown. When dry the leaves curl in a haphazard manner and give the plant a tousled appearance. Up to 3mm long, the narrow, tapering leaves have an extended leaf-base, around the edges of which is an easily observed band. The tapered part of the leaf possesses a keel and the nerve disappears into the pointed apex. The capsule is erect and, at first, is covered by a hairy calyptra. When the calyptra has been shed the capsule displays eight prominent longitudinal ribs, a narrow neck and a constriction near the mouth. The capsule is about 2mm long and the seta between 10 and 12mm.

Hookeria lucens

The nearest relatives of this rather leafy-looking moss are tropical and it was once thought that it was very rare. However, it is now known to be fairly common in suitable habitats. The large leaves are about 5mm long and the thin-walled cells, which can be seen with the naked eye, give it a filmy translucence. There are no serrations on the leaf margin and the central nerve is likewise absent. The dark, purplish-black capsule is held horizontally or slightly drooping on a blackish seta about 2cm long. The leaves are spaced along branches 2 to 6cm long — the whole possessing a flattened appearance. *Hookeria lucens* requires deep shade and moist conditions. It is usually found on soil, either on rock ledges or in moist woodland. The specimen drawn was collected in Glengarriff, Co. Cork.

21 Algae/Seaweeds

The Brown Algae *Phaeophyceae*

The algae represent some of the simplest forms of plant life. A large number consist of single cells only and are microscopic in size. Many are of immense importance in their ecological significance, although their size limits their impact on most people. The brown algae, however, consists of well-known seaweeds — some species being amongst the 'longest' of living organisms. In the Antarctic, *Macrocystis pyrifera,* a giant species of oarweed, attains a length of 70m! There are few rocky shores in the world where brown seaweeds do not occur. Their characteristic colour is due to the presence of the pigment Fucoxanthin, which masks the green of the photosynthetic pigment chlorophyll. Only a handful of species occur in freshwater, otherwise they are exclusively marine.

Peacock's Tail *Padina pavonia*
In the family DICTYOTACEAE is classified the unusual Peacock's tail seaweed. There is first a minute disc which becomes coated with brown fibres. A slender, round stem then grows laterally to produce a broad lamina, increasing gradually in width and then expanding into a thin fan-shaped frond up to 10cm wide. During subsequent growth each frond becomes concave and often funnel-shaped. The outer-surface becomes banded with light and dark brown and olive-green. An alternation of generations occurs in which a tetraspore-producing generation follows a male and female gamete-producing generation. *Padina* is essentially tropical and subtropical in distribution. It has been found only once or twice on the south coast of Ireland.

The Wrack Family *Fucaceae*
Some of the well-known and larger brown seaweeds belong to the 'wracks'. On rocky shores throughout the north temperate region, they cannot fail to be noticed, even by the least observant of persons. Mostly they are 'short-lived' perennials. It has been estimated that more than a half of the plants under three years of age are torn up by winter storms. In general form there is a disc-like holdfast for anchorage to rocks and from this spring the ribbon-like fronds which divide repeatedly in the same plane. The ends of the fronds are thickened and are known as the 'receptacles' and these contain the sexual organs. Antheridia and oogonia are produced in small pits on the surface of the receptacles. Some plants produce both male and female receptacles whilst others are on different plants. The different species of wrack form, more or less, distinct zones at the various tide levels. There is no vegetative reproduction,

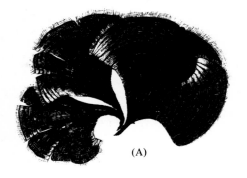

(A)

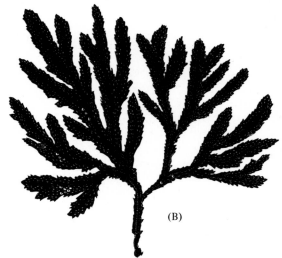

(B)

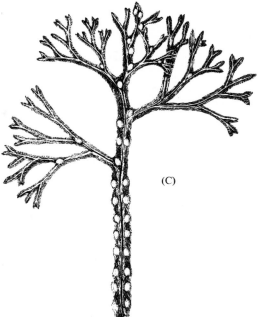

(C)

(a) Peacock's Tail, *Padina pavonia*. Frond up to 10 cm in width. Rare in the South in sunny rock pools.

(b) Toothed Wrack, *Fucus serratus*. Frond segments are about 18 mm in width and total length of plant about 60 cm.

(c) Bladder Wrack, *Fucus vesiculosus*. Total length of plant varies from 30 cm to 1 m. Very common.

(d) Spiral Wrack, *Fucus spiralis*. Total length of plant varies from 15-35 cm. Common.

(e) Horned Wrack, *Fucus ceranoides*. Total length from 25-45 cm. In brackish water and land-locked bays. Not usually with other *Fucus* species.

(E)

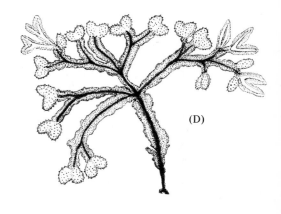

(D)

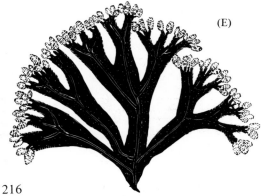

216

although in a number of species injuries will cause rejuvenation — shown by a proliferation of new growth.

The Toothed Wrack *Fucus serratus*

This is one of the most abundant and most easily identified of the wracks. It forms a dense growth on rocky shores just above or near low-tide level. The olive-brown fronds are margined by projecting teeth and there is a thick, conspicuous midrib. The flat frond segments are about 18mm in width and the total length of the plant is about 60cm. The surface of the frond segment is covered with minute pin-prick-like cavities, rendered noticeable by colourless hairs which emerge from them.

The receptacles show little differentiation and are found at the frond ends — slightly thickened patches up to about 5cm in length, but the male and female receptacles are on different plants, with the orange colour of the ripe male receptacle making identification simple. Although predominantly fertile during winter, fruiting plants may be found at most times. A characteristic of this species is the presence often of a small worm *Spirorbis*, which constructs a white, spiral shell about 2 to 3mm in diameter and is fastened to the frond — often in considerable numbers.

Bladder Wrack *Fucus vesiculosus*

This is an extremely variable species in form and size — so much so that difficulties may sometimes be encountered in its identification. Its usual habitat is the middle region of the shore and usually above the Toothed wrack zone, in semi-sheltered situations. The typical form occurs in these areas.

Numbers of pea-sized air vesicles occur — generally in pairs — on either side of the midrib of the fronds. These act as floats and lift the fronds with the incoming tide. The frond-segments are not toothed but are somewhat sinuate, due to the presence of the vesicles. The length of this plant is from 30cm to over 1m, and repeated forking is shown. Male and female receptacles occur on different plants and are bright orange in the former and yellowish-olive

in the latter. On exposed shores, or when growing on a steeply-sloping rock face, the fronds are very narrow and vesicles often absent, with the plant dwarfed. In very sheltered situations the vesicles may be so abundant that they almost touch each other.

Spiral Wrack *Fucus spiralis*

Spiral wrack is relatively small, being from 15 to 35cm in length. The fronds are sometimes, but, by no means always, twisted — hence its specific name. However, this attribute is most unreliable in identification. The holdfast is disc-shaped and the segments of the frond are strap-shaped and are about 12mm wide with a prominent midrib. The margins of the frond-segment are irregularly sinuate. Minute pits cover the frond and are especially abundant in the region of the growing tip.

The receptacles are light in colour, ellipsoidal and up to about 16mm in width and often bifurcate, as in the specimen illustrated. The swollen receptacles are said to resemble 'clusters of large sultanas', and they are hermaphrodite. Some specimens of this plant have large blisters on the fronds which could cause confusion with Bladder wrack. Spiral wrack grows on the upper beach zone, above those of Bladder wrack and *Ascophyllum*. The receptacles, known as 'jelly bags', were used in hot, salt water as a cure for corns in some parts of Ireland.

Horned Wrack *Fucus ceranoides*

The Horned wrack is relatively small, being only from 25 to 45cm in length. The frond segments are up to 12mm in width, with untoothed margins which may, however, be slightly undulate. The midrib is prominent and the lower laminae (lateral extensions of the fronds) often disappear. The lateral frond segments are not nearly as wide as the main axis.

In colour it is a pale-greenish olive and the receptacles are in fan-shaped groups. Usually these are long and pointed, but the specimen illustrated (which is atypical) shows them to be ellipsoidal. Although usually both sexes are found on the same plant, sometimes only one sex occurs. Its usual habitat is in brackish water in

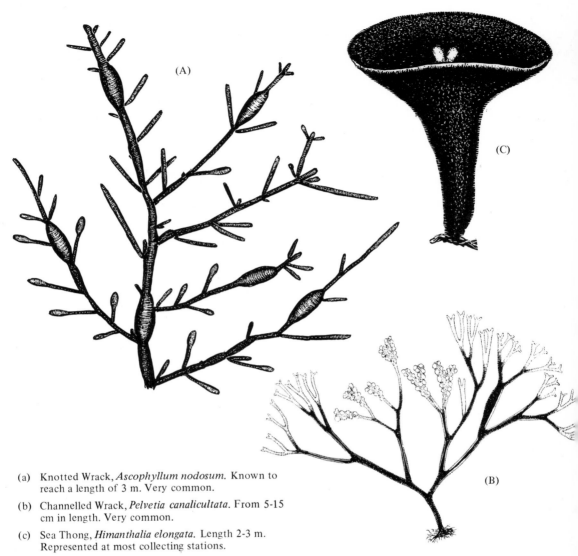

(A)

(C)

(B)

(a) Knotted Wrack, *Ascophyllum nodosum.* Known to
     reach a length of 3 m. Very common.

(b) Channelled Wrack, *Pelvetia canalicultata.* From 5-15
     cm in length. Very common.

(c) Sea Thong, *Himanthalia elongata.* Length 2-3 m.
     Represented at most collecting stations.

(d) Irish Moss, *Chondrus crispus.* Up to 22 cm in length
     but usually only about half this size. Particularly
     common around the Irish coast. Two types of frond.

(D)

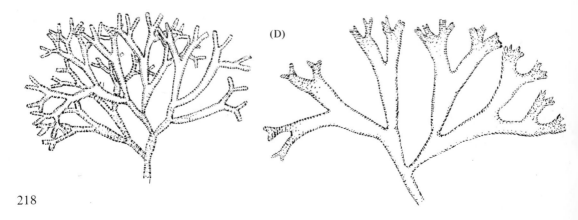

estuaries or land-locked bays. The specimen illustrated may have originated near the inflowing river of Dunmanus Bay, eventually being washed out towards the mouth of the Bay.

Knotted Wrack *Ascophyllum nodosum*
Knotted wrack is identified by the absence of midribs and the minute surface pits. The basal part of the main frond is rounded for some length before becoming flattened. Large air-bladders are produced – never in pairs – at intervals of about 5 to 10cm, which may vary in shape from spherical to ellipsoidal. One bladder is produced each year, allowing the age of the plant to be determined. At shorter intervals, small branches are produced which end in receptacles. In the male these are golden yellow whilst those of the female are yellowish-olive. This plant is known to reach a length of 3m in favourable conditions. Knotted Wrack generally grows in sheltered bays where the shore shelves gently and where it can obtain a good anchorage for its holdfast. It is usually found above the Serrated wracks' zone.

Channelled Wrack *Pelvetia canalicultata*
This species of wrack is separated from the various *Fucus* species by having no midrib, and from the Knotted wrack by having no bladders. It grows in short tufts from 5 to 15cm in length from high-water mark down to the Spiral wrack zone. The fronds are identified by their inrolled margins (channels), and are about 6mm in width. The long receptacles are forked and the hermaphrodite sexual bodies are conspicuous, being a yellowish-green in colour.

The channelling of the fronds forms an important function. Occupying the highest of the brown seaweed zones, this plant spends much time out of water – indeed, in some localities it even grows above high-water mark and is only in a 'splash' zone. In such unfavourable conditions it lies flat against the rock surface, retaining essential amounts of sea-water in its channels. Sheep and cattle will feed on a number of seaweed species, including this one.

Sea Thong *Himanthalia elongata*
The Sea thong is classified in the family HIMANTHALIACEAE. It is one of the brown seaweeds which is easy to identify, for when young, this plant consists of a conical, button-shaped body with a concave upper surface. It is about 2.5cm in diameter and has a polished texture, and is olive-brown in colour. From the centre of the concavity emerge several fertile branches which grow to a length of 2 to 3m and are yellowish-brown in colour. These are from 3 to 10mm in width and fork repeatedly. The 'thongs' bear sexual organs and the sexes are born on different plants. This species occurs below the Spiral wrack zone and is often found amongst the Oarweeds *(Laminaria)*.

Irish Moss (Carragheen) *Chondrus crispus*
The RHODOPHYTA consists of a number of families of 'Red' seaweeds which have a simple organization and which have either thin membraneous, gelatinous fronds, or are filamentous. Many of them are small. *Chondus crispus,* together with other species of the family GIGARTINACEAE is popularly known as Irish Moss or Carragheen. It is purplish-red in colour (in sunlit pools it may be green) and averages about 7 to 8cm in length. It has a disc-shaped holdfast which adheres to a stone, and there may be as many as twenty fronds springing from it. The narrow base of the frond expands to a width of from 12 to 20cm and branches out dichotomously from six to eight times. There is, however, great variation in frond width. This species occurs in stony situations in pools from the lower shore into the Oarweed zone.

Carragheen has been well-known as a medicine for nearly 150 years, being used principally for lung complaints. It is still used (to a limited extent) in the manufacture of jellies. *Gigartina stellata* resembles *Chondrus crispus* to a great extent and is often associated with it. When mature, *G.stellata* produces large numbers of papillae which give it a ragged appearance, but *C.crispus* does not do this. It has been widely used as a substitute for Japanese agar for culture of bacteria. It occurs mostly at about the level of low spring-tides. My seaweeds were identified by Miss M. Scannell, Keeper of the Herbarium at the National Botanic Gardens at Glasnevin, Dublin 9.

General Index

Bold Type indicates principal entries in book.
see Geographical Index for all references to places and
List of Main Entries for English/Latin/Irish names.

227

Geographical Index

List of Main Entries

This is a listing of all the principal entries in this book, the vast majority of which are illustrated; the animals and plants are listed separately.

Animals

LATIN NAME	ENGLISH NAME	IRISH NAME
Accipiter nisus	Sparrowhawk	Spioróg
Acherontia atropus	Death's head hawk-moth	Conach na cealtrach
Acipenser sturio	Sturgeon	Bradán fearna
Adalia bipunctata	Two-spot ladybird	Bóin Dé
Adalia 10-punctata	Ten-spot ladybird	Bóin Dé
Aegeria scoliaeformis	Welsh clearwing moth	Leamhan
Aeolidia papillosa	Plumed aeolis	Bodalach
Agelina labyrinthica	Gorse-web spider	Damhán alla
Aglais urticae	Small tortoiseshell	Ruán beag
Alopius vulpinus	Thresher	Sionnach mara
Alosa fallax killarnensis	Killarney shad	Sead fhallacsach
Ammodytes tobianus	Lesser sand-eel	Corr ghainimh
Anobium punctatum	Common woodworm	Míol críon
Aphrodite aculeata	Sea mouse	Luch mhara

LATIN NAME	ENGLISH NAME	IRISH NAME
Aplysia punctata	Sea hare	– – –
Apodemus sylvaticus	Wood mouse	Luch fhéir
Apus apus	Swift	Gabhlán gaoithe
Argynnis paphia	Silver-washed fritillary	Fritileán geal
Asterias rubens	Common starfish	Crosóg mhara
Asterina gibbosa	Common starlet	Crosán Faoileáin
Atelecyclus septemdentatus	– – –	Portán cruinn
Aythya fuligula	Tufted duck	Lacha bhadánach
Balaena glacialis	Biscayan whale	Míol mór
Balaenoptera acutorostrata	Lesser rorqual	Droimeiteach beag
Balaenoptera borealis	Sei whale	Míol mór
Balaenoptera musculus	Blue whale	Míol mór gorm
Balaenoptera physalus	Common rorqual	Droimeiteach
Blennius gattorugine	Tompot blenny	Cloigeann chruachain
Brachydesmus superus	– – –	Mílechosach
Brachygeophilus truncorum	– – –	Céadchosach
Branta canadensis	Canada goose	Gé
Bufo calamitas	Kerry toad	Cnádán
Calamia tridens ssp. occidentalis	Burren green moth	Leamhan
Callionymus lyra	Dragonet	Iascán nimhe
Callochiton achatinus	Smooth chiton	Ciotón
Callophhrys rubi	Green hairstreak butterfly	Stiallach uaine
Capra hircus	Wild goat	Gabhar fia
Caprella aequilibra	Skeleton shrimp	Séacla
Carabus clathratus	Clathrate ground beetle	Daol
Cardium echinatum	Prickly cockle	Ruacan garbh
Cervus elephas	Red deer	Fia rua
Cervus nippon	Sika deer	Fia seapánach
Cetonia aurata	Rose chafer	Deá
Cetorhinus maximus	Basking shark	Liamhán gréine
Cilix glaucata	Chinese character moth	Leamhan
Cinclus cinclus hibernicus	Irish dipper	Gabha dubh
Circus cyaneus	Hen harrier	Croman na gcearc
Clethrionomys glareolus	Bank vole	Vól bruaigh
Coenonympha pamphilus	Small heath butterfly	Fraochán beag
Conilera cylindracea	– – –	Cláirseach
Corvus corone cornix	Hooded crow	Feannóg
Corvus frugilegus	Rook	Rúcach
Corvus monedula	Jackdaw	Cág
Corystes cassivelaunus	Masked crab	Portán clismín
Crex crex	Corncrake	Traonach
Cryphia muralis ssp. westroppi	Marbled green moth	Leamhan
Cycnia mendica ssp. rustica	Muslin moth	Leamhan
Cygnus olor	Mute swan	Eala bhalbh
Dama dama	Fallow deer	Fia buí
Delphinus delphis	Common dolphin	Deilf
Doris tuberculata	Rough sea lemon	Bodalach
Dytiscus marginalis	Great diving beetle	Tumadóir mór

LATIN NAME	ENGLISH NAME	IRISH NAME
Echinocardium cordatum	Common heart urchin	Madra mór
Echinus esculentus	Edible sea urchin	Cuán mara
Enoplognatha ovata	Oval bush spider	Damhán alla
Ensis silliqua	Pod razor	Scian mhara
Equus caballus	Connemara pony	Capaillín Chonamara
Erinaceus europaeus	Hedgehog	Gráinneog
Erynnis tages	Dingy skipper	Donnán
Esox lucius	Pike	Liús
Euchloë cardamines ssp. hibernica	Orange-tip butterfly	Barr buí
Eupatia bilineata ssp. isolata/ hibernica	Yellow shell moth	Leamhan
Euphydryas aurinea ssp. hibernica	Marsh fritillary	Fritileán réisc
Eupithecia venosata ab. plumbea	Netted pug moth	Leamhan
Eurynebria complanata	―――	―――
Falco tinnunculus	Kestrel	Pocaire gaoithe
Gadus morhua	Cod	Trosc
Gallinago gallinago	Snipe	Naoscach
Geomalacus maculosus	Kerry slug	―――
Globicephala melaena	Pilot whale	Píolótach
Gobius paganallus	Rock goby	Mac siobháin carraige
Grampus griseus	Risso's dolphin	Deilf
Hadena caesia	Grey moth	Leamhan
Hadena lepida ssp. capsophila	Pod lover moth	Leamhan
Haedropleura septangularis	Seven-ribbed conelet	―――
Halichoerus grypus	Grey seal	Rón glas
Helix nemoralis	Banded snail	Seilide
Heterosepiola atlantica	Little cuttle fish	Cudal
Hirudo medicinalis	Medicinal leech	Súmaire fola
Hiatella arctica	Wrinkled rock-borer	―――
Holothuria forskali	Sea cucumber	Súmaire cladaigh
Hyperoodon ampullatus	Bottle-nosed whale	Míol mór bolgshrónach
Issus coleoptratus	Irish issus	―――
Lacerta vivipara	Viviparous lizard	Earc luachra
Laothoe populi	Poplar hawk-moth	Conach poibleoige
Lagenorhynchus acutus	White-sided dolphin	―――
Lagenorhynchus albirostris	White-beaked dolphin	―――
Larvaevora grossa	Great black parasite fly	―――
Lepadogaster candollei	Connemara sucker	Leatha leice
Lepas anatifera	Goose barnacle	Giúrann gasach
Leptidea sinapis ssp. juvernica	Wood white	Bánóg choille
Lepus capensis	Brown hare	Giorria gallda
Lepus timidus hibernicus	Irish hare	Giorria
Leucodonta bicoloria	White prominent (moth)	Starraicín
Limnephilus fuscinervis	Brown-veined limnephilus	Cuil chadáin
Liparis montagui	Montagu's sea snail	Gnamhán
Lithobius forficatus	―――	Céadchosach cré
Loligo forbesii	Common squid	Máthair shúigh

LATIN NAME	ENGLISH NAME	IRISH NAME
Lophius piscatorius	Angler fish	Láimhíneach
Luperina nickerlii ssp. knilli	Sandhill rustic moth	Leamhan
Lutra lutra	Otter	Madra uisce
Machilis (Petrobius) maritima	Maritime bristletail	Earr ghuaireach
Maniola jurtina ssp. iernes	Meadow brown butterfly	Donnóg fhéir
Martes martes	Pine marten	Cat crainn
Megaloceros giganteus	Irish giant deer	Fia mór na mbeann
Megaptera novaeangliae	Humpback whale	Míol mór
Meles meles	Badger	Broc
Mesoplodon bidens	Sowerby's whale	Míol mór
Mesoplodon mirus	True's beaked whale	Míol mór
Mitopus morio	Harvestman	Pilib an fhómhair
Mola mola	Sunfish	Iasc na gréine
Molva molva	Ling	Langa
Motacilla alba	Pied wagtail	Glasóg shráide
Mus musculus	House mouse	Luch thí
Mustela ermines hibernica	Irish stoat	Easóg
Mustela vison	American mink	Minc Mheiriceánach
Myotis daubentoni	Daubenton's bat	Ialtóg Dhaubenton
Myotis mystacinus	Whiskered bat	Ialtóg ghiobach
Myotis nattereri	Natterer's bat	Ialtóg Natterer
Nephropa norvegicus	Dublin Bay prawn	– – –
Neptunea antiqua	Red whelk or Buckie	– – –
Numenius arquata	Curlew	Crotach
Nyctalus leisleri	Leisler's bat	Ialtóg Leisler
Nymphalis io	Peacock butterfly	Péacóg
Ophiothrix fragilis	Common brittlestar	Crosóg bhriosc
Opisthograptis luteolata	Brimstone moth	Leamhan ruibheach
Orcinus orca	Killer whale	Cráin dhubh
Oryctolagus cuniculus	Rabbit	Coinín
Otiorrhynchus ruropunctatus	Red-spotted weevil	Gobachán
Pagurus bernhardus	Hermit crab	Faocha ghilomaigh
Palmipes membranaceus	Duck's foot starfish	Crosóg mhara
Paracentrotus lividus	Purple sea urchin	Cuán mara deilgneach
Pararge aegeria	Speckled wood butterfly	Breacfhéileacán coille
Patella vulgata	Common limpet	Bairneach
Pecten maximus	Great scallop	Muirín
Phalacrocorax carbo	Cormorant	Broigheall
Philudoria potatoria	Drinker moth	Leamhan
Phoca vitulina	Common seal	Rón beag
Phocoena phocoena	Common porpoise	Muc mhara
Pholadides loscombiana	Paper piddock	– – –
Pholcus phalangioides	– – –	– – –
Physeter catodon	Sperm whale	Caisealóid
Pieris napi	Green-veined white butterfly	Bánóg uaine
Pipistrellus pipistrellus	Pipistrelle bat	Ialtóg fheascrach
Plecotus auritus	Long-eared bat	Ialtóg chluasach
Podiceps cristatus	Great crested grebe	Foitheach mór

LATIN NAME	ENGLISH NAME	IRISH NAME
Porania pulvillus	Cushion starfish	Crosóg mhara
Proneomenia aglaopheniae	— —	— — —
Propylea 14-punctata	Fourteen-spot ladybird	Bóin Dé
Pseudopanthera macularia	Speckled yellow moth	Leamhan
Pyronia tithonus	Gatekeeper butterfly	Geatóir
Pyrrhocorax pyrrhocorax	Chough	Cág cosdearg
Raja clavata	Thornback ray	Roc garbh
Rana temporaria	Frog	Frog
Rattus norvegicus	Brown rat	Francach donn
Rattus rattus	Black rat	Francach dubh
Rhagium inquisitor	Inquisitive longhorn	Ciaróg fhadadharcach
Rheumaptera hastata	Argent and sable moth	Leamhan
Rhincophus hipposideros	Lesser horse-shoe bat	Crú-ialtóg bheag
Salvelinus alpinus	Char	Ruabhreac
Saxicola torquata	Stonechat	Caislín cloch
Sciurus carolinensis	American grey squirrel	Lora glas
Sciurus vulgaris	Red squirrel	Lora rua
Scolopax rusticola	Woodcock	Creabhar
Sepia elegans	Elegant cuttle fish	Cudal
Setina irrorella	Dew moth	Leamhan
Sigara fallenoidea	Irish water boatman	Bádóir
Smerinthus ocellata	Eyed hawk-moth	Seabhacleamhan
Soaphander lignaria	Canoe shell	— — —
Solaster papposus	Common sunstar	Crosóg ghréine
Somatochlora arctica	Northern emerald dragonfly	Snáthaid mhór
Sorex minutus	Pygmy shrew	Dallóg fhraoigh
Spirorbis borealis	— — —	Péist na bhféadán
Sterna albifrons	Little tern	Geabhróg bheag
Streptopelia decaocto	Collared dove	Fearán baicdhubh
Syngnathus acus	Pipefish	Snáthaid mhara
Terebella lapidaria	— — —	Péist ghuaireach
Teredo norvagica	Shipworm	Rincs
Tethea or ssp. hibernica	Poplar lutestring moth	Leamhan
Tinca tinca	Tench	Cúramán
Tinodes assimilis	Wet rock tinodes	Cuil chadáin
Tipula maxima	Giant crane fly	Galán
Tricla lignaria	Canoe shell	— — —
Trigloporus lastoviza	Streaked gurnard	Cnúdán breac
Tringa hypoleucos	Common sandpiper	Gobadán
Triturus vulgaris	Common newt	Earc sléibhe
Trivia monacha	Cowrie	Fínicín
Troglodytes troglodytes	Wren	Dreolín
Tursiops truncatus	Bottle-nosed dolphin	Deilf
Upogebia deltaura	Burrowing shrimp	— — —
Urocerus gigas	Giant wood wasp	Sábhchuil
Velella velella	By-the-wind-sailor	— — —
Volucella pellucens	White-banded drone fly	Beach ghabhair

LATIN NAME	ENGLISH NAME	IRISH NAME
Vulpes vulpes	Fox	Sionnach
Xysticus cristatus	Crab spider	Damhán alla
Zeus faber	John Dory	Deoraí
Ziphius cavirostris	Cuvier's whale	Míol mór Chuvier
Zygaena purpuralis	Transparent burnet moth	Buirnéad

Plants

LATIN NAME	ENGLISH NAME	IRISH NAME
Acer campestre	Field maple	Mailp
Adiantum capillus-veneris	Maidenhair fern	Dúchosach
Allium triquetrum	Three-cornered leek or Triquetrous garlic	Gairleog
Amanita pantherina	Leopard amanita	— — —
Ammophila arenaria	Marram grass	Muiríneach
Anagallis arvensis	Scarlet pimpernel	Falcaire fiáin
Antennaria dioica	Mountain cudweed	Catluibh
Anthoxanthum odoratum	Vernal grass	Féar cumhra
Arbutus unedo	Strawberry tree	Caithne
Armeria maritima	Thrift plant	Rabhán
Arrhenatherum elatius	False oat	Coirce bréige
Arum maculatum	Lords and ladies	Cluas chaoin
Ascophyllum nodosum	Knotted wrack	Feamainn bhuí
Asplenium adiantum-nigrum	Black spleenwort	Fionncha dubh
Asplenium trichomanes	Maidenhair spleenwort	Fionncha dúchosach
Aster tripolium	Sea aster	Luibh bhléine
Blechnum spicant	Hard fern	Raithneach chrua
Boletus edulis	Penny bun	— — —
Botrychium lunaria	Moonwort	Lus na míosa
Bryum argenteum	— — —	Caonach
Calystegia soldanella	Sea bindweed	Ialus
Cantherrellus cibarius	Chanterelle	Cantarnaid
Carex acutiformis	Lesser pond sedge	Cíb
Carex echinata	Star sedge	Cíb réalta
Ceterach officinarum	Rusty-back fern	Raithneach rua
Chondrus crispus	Irish moss (Carragheen)	Cosáinín (carraige)
Cochlearia officinalis	Scurvy-grass	Biolar trá
Coprinus atramentarius	Common ink-cap	— — —
Coriolus versicolor	— — —	— — —
Cuscuta epithymum	Dodder	Clamhán
Cynosurus cristatus	Dog's tail	Coinfhéar
Daboecia cantabrica	St Daboec's heath	Fraoch na haon choise

LATIN NAME	ENGLISH NAME	IRISH NAME
Dactylorhiza majalis	Broad-leaved marsh orchid	———
Dicranella heteromalla	———	Gabhalchaonach
Dicranum major	———	Gabhalchaonach
Drosera rotundifolia	Common sundew	Drúchtin móna
Dryas octopetala	Mountain avens	Féasóg na lao
Elymus farctus	Sand couch	———
Epilobium nerterioides	New Zealand willow-herb	Saileachán
Erica erigena	Mediterranean heath	Fraoch camógach
Erica tetralix mackaiana	Mackay's heath	Fraoch naoscaí
Erigeron mucronatus	Mexicane fleabane	———
Eryngium maritimum	Sea holly	Cuileann trá
Escallonia macrantha	———	———
Euphorbia hyberna	Irish spurge	Bainne caoin
Fuchsia magellanica or gracilis	———	Fiúise
Fucus ceranoides	Horned wrack	Feamainn
Fucus serratus	Toothed wrack	Míoránach
Fucus spiralis	Spiral wrack	Casfheamainn
Fucus vesiculosus	Bladder wrack	Feamainn bhoilgíneach
Gentiana verna	Spring gentian	Ceadharlach Bealtaine
Glaucium flavum	Horned poppy	Caillichín na trá
Hesperis matronalis	Dame's violet	———
Himanthalia elongata	Sea thong	Ríseach
Holcus lanatus	Yorkshire fog	Féar an chinn bháin
Hookeria lucens	———	Caonach
Hordeum murinum	Wall barley	Cuiseogach fhionn
Hygrophorus conicus	———	———
Hymenophyllum spp.	Filmy fern	Raithneach scannánach
Hypericum androsaemum	Tutsan	Meas torc allta
Hypericum elodes	Marsh St John's wort	Beathnua corraigh
Hypericum humifusum	Trailing St John's wort	Beathnua
Iris pseudacorus	Yellow flag	Feileastram
Jasione montana	Sheep's bit scabious	Duán na gcaorach
Juniperus communis	Juniper	Aiteal
Lactarius blennius	———	———
Larix leptolepsis	Japanese larch	Learóg Sheapánach
Lavatera arborea	Tree mallow	Leamhach chrainn
Lepiota procera	Parasol mushroom	———
Leucobryum glaucum	———	Gabhalchaonach
Lychnis flos-cuculi	Ragged robin	Lus síoda
Lycoperdon perlatum	Puff ball	Bolgán béice
Lysimachia nemorum	Yellow pimpernel	Lus Cholm Cille
Meconopsis cambrica	Welsh poppy	Poipín caimbriach
Mnium punctatum	———	Snáthchaonach

LATIN NAME	ENGLISH NAME	IRISH NAME
Narthecium ossifragum	Bog asphodel	Sciollam na móna
Ophioglossum lusitanicum	Dwarf adder's tongue	———
Ophioglossum vulgatum	Adder's tongue fern	Lus na teanga
Ophrys apifera	Bee orchid	Magairlín beachach
Ophrys insectifera	Fly orchid	Magairlín na cuileoige
Orchis mascula	Early purple orchis	Magairlín meidhreach
Osmunda regalis	Royal fern	Raithneach ríúil
Oxalis acetosella	Wood sorrel	Seamsóg
Padina pavonia	Peacock's tail	———
Pelvetia canalicultata	Channelled wrack	Caisíneach
Petasites fragrans	Winter heliotrope	Plúr na gréine
Petasites hybridus	Butterbur	Gallán mór
Phyllitis scolopendrium	Hart's tongue fern	Creamh na muice fia
Picea sitchensis	Sitka spruce	Sprús Sitceach
Pinguicula grandiflora	Kerry violet	Sadhbhóg shléibhe
Pinguicula lusitanica	Pale butterwort	Leith uisce
Pinus sylvestris	Scots pine	Péine Albanach
Plantago coronopus	Buck's horn plantain	Adharca fia
Polypodium vulgare	Polypody	Scim choiteann
Polyporus squamosus	Dryad's saddle	Bracfhungas
Polytrichum juniperinum	———	Foltchaonach
Potentilla fruticosa	Shrubby cinquefoil	———
Psalliota arvensis	Horse mushroom	Agairg
Pseudotsuga taxifolia	Douglas fir	Giúis Dhúghlais
Pteridium aquilinium	Bracken	Raithneach mhór
Ptychomitrium polyphyllum	———	———
Ranunculus flammula	Lesser spearwort	Glassair léana bheag
Rannunculus omiophyllus	Moorland crowfoot	Crobh préacháin
Rubus saxitilis	Stone bramble	Sú na mban mín
Salix repens	Creeping willow	Saileach reatha
Saxifraga granulata	Meadow saxifrage	Mionán Muire
Saxifraga hirsuta	Kidney saxifrage	Mionán duánach
Saxifraga spathularis	St Patrick's Cabbage or London Pride	Cabáiste an mhadra rua
Scirpus caespitosus	Tufted sedge	Cíb cheanngheal
Scleroderma verrucosus	Warty earth-ball	———
Scutellaria minor	Lesser skullcap	Cochall
Sedum anglicum	English stonecrop	Póiríní seangán
Silene vulgaris ssp. maritima	Sea campion	Coireán cuile
Simethis planifolia	Kerry lily	Lile Chiarraíoch
Spiranthes spiralis	Ladies' tresses	Cúilín Mhuire
Stellaria holostea	Greater stitchwort	Tursarraing mhór
Taxus baccata	Yew	Iúr
Triglochin maritima	Sea arrow-grass	Barr an mhilltigh mara
Ubilicus rupestris	Pennywort	Cornán caisil
Ulota crispa	———	———

LATIN NAME	ENGLISH NAME	IRISH NAME
Viola palustris	Marsh violet	Sailchuach chorraigh
Viola tricolor	Wild pansy	Goirmín searraigh
Wahlenbergia hederacea	Ivy-leaved bellflower	Clogbhláth eidhneánach

For Further Reading

The following list of references can only be a cross-section of the literature on Irish fauna and flora. The majority of the titles are in print and are currently available from booksellers, publishing houses or the cited institutions. Some of them are inexpensive and their acquisition would make an excellent beginning for a library on the wildlife of Ireland. The list is organised according to the author-date system. **N.D.** indicates that a book contains no date of publication.

ANON, *Report of the Minister for Lands.* Forest and Wildlife Service, April 1973 to Dec. 1974. Stationery Office, Dublin.

ANON, N.D. *Wetlands Discovered.* Department of Lands, Forest and Wildlife Service.Stationery Office, Dublin.

ARMSTRONG, E. A., 1955, *The Wren,* Collins, London.

AUSTIN, O., 1961. *Birds of the World.* Paul Hamlyn, London.

BARNES, R. D. 1974 (3rd edition). *Invertebrate Zoology.* Saunders, Philadelphia.

BARBER, P. and PHILLIPS, C. E. L., 1975. *The Trees around us.* Weidenfeld and Nicolson, London.

BAYNES, F. S. A., 1964 and revised edition 1973. *A Revised Catalogue of Irish Macrolepidoptera.* Classey, Hampton. Middlesex.

BAYNES, F. S. A. 1970. *Supplement to a Revised Catalogue of Irish Macrolepidoptera.* Classey, Hampton. Middlesex.

BEAN, W. J., 1914 and other editions. *Trees and Shrubs Hardy in the British Isles.* Murray, London.

BENTHAM, G., and HOOKER, Sir J. D., 1930 (7th Edition). *Handbook of the British Flora.* Reeve, Ashford, Kent, England.

CLAPHAM, A. R., TUTIN, T. G., and WARBURG, E. F., 1958. *Flora of the British Isles.* Cambridge University Press.

CLARK, A. M., 1968 (2nd Edition). *Starfishes and their Relations.* British Museum (Natural History), London.

CLOUDSLEY-THOMPSON, J. L., and SANKEY, J., 1961 reprinted 1968. *Land Invertebrates.* Methuen, London.

COLYER, C. N., and HAMMOND, C. O., 1951. *Flies of the British Isles.* Warne, London.

COOPE, G. R., 1973. "The Ancient World of 'Megaceros'." *Deer* 2 (10): 974-977.

COOPE, G. R., 1968. "The Evolutionary Origin of Antlers." *Deer* 1: 215-217.

COWDY, S., 1976. "A Cry on the Wind." *Birds,* the RSPB magazine. Autumn 1976; 31-32 and front cover illustration by John Busby.

CULLINANE, J. P., 1973. *Phycology of the South Coast of Ireland.* Cork University Press, Cork.

DIXON, P. S., and IRVINE, L. M., 1977. *Seaweeds of the British Isles.* Vol. I, *Rhodophyta,* Part 1. British Museum (Natural History), London.

DONALDSON, F., N.D. *The Lusitanean Flora.* Irish Environmental Series. Folens, Dublin.

EASON, E. H., 1964. *Centipedes of the British Isles.* Warne, London.

ETTLINGER, D. M. T., 1976. *British and Irish Orchids – A Field Guide.* Macmillan Press, London.

FAIRLEY, J. S., 1975 *An Irish Beast Book.* Blackstaff Press, Belfast.

FINDLAY, W. P. K., 1967. *Wayside and Woodland Fungi.* Warne, London.

FITTER, R., and FITTER, A., 1974. *The Wild Flowers of Britain and Northern Europe.* Collins, London.

FRASER, F. C., 1976 (5th Edition). *British Whales, Dolphins and Porpoises.* British Museum (Natural History), London.

HAMMOND, C. O., 1977. *The Dragonflies of Great Britain and Ireland.* Curiven Press, London.

HOWARTH, T. G., 1973. *South's British Butterflies.* Warne, London.

HUTCHINSON, C., editor, 1975. *The Birds of Dublin and Wicklow,* Irish Wildbird Conservancy, Dublin.

JOHNSON & HALBERT, 1902. "A List of the Beetles of Ireland". *Proc. R. Irish Acad.,* Dublin. 6(4): 535-827.

KEMP, S. W., 1901-04, "The Beetles of the Limerick District", *Journal, Limerick Field Club,* 2: 269-279.

LINSSEN, E., 1959. *Beetles of the British Isles.* 2 Vols. Warne, London.

McMILLAN, N. F., 1968, *British Shells,* Warne, London.

McNALLY, K., 1976. *The Sun-fish Hunt,* Blackstaff Press, Belfast.

MANN, K. H. & WATSON, E. V. 1954. *A Key to the British Freshwater Leeches with Notes on their Ecology.* Scientific Publications No. 14. Freshwater Biological Association, Amblende.

MARSHALL, N. B., 1971, *Ocean Life in Colour.* Blandford, London.

MEGLITSCH, P. A. 1972 (2nd edition). *Invertebrate Zoology.* Oxford University Press, New York.

MORIARTY, C., 1967. *A Guide to Irish Birds.* Mercier Press, Cork.

MORIARTY, C., N.D. *A Natural History of Ireland.* Mercier Press, Cork.

MUUS, B. J., 1974 (1964 Danish edition). *Collins Guide to the Sea Fishes of Britain and North-western Europe.* (Translated by Vevers, G.) Collins, London.

NASH, R., 1975. "The Butterflies of Ireland". *Soc. Brit. Ent. Nat. Hist.* 7 (3): 69-73.

NELSON, E. C., and BRADY, A., (Eds), 1979. *Irish Gardening and Horticulture* Royal Horticulture Society of Ireland, Dublin.

Ó CÉIDIGH, 1963. *A List of Irish Marine Decapod Crustacea.* Stationery Office, Dublin.

O'CONNELL, G., (Editor) N.D. *The Burren – A Guide.* Shannonside, Mid-Western Regional Tourism Organization Ltd., Limerick.

O'MAHONY, E., 1929. (In Praeger, R. L., Report on recent additions to the Irish Fauna and Flora). *Proc. R. Irish Acad.* 39(B): 22-94 Dublin.

O'RIORDAIN, C. E., 1965. *A Catalogue of the Collection of Irish Fishes in the National Museum of Ireland.* Stationery Office, Dublin.

O'RIORDAIN, C. E., 1972. "Provisional List of Cetacea and Turtles stranded or captured on the Irish Coast". Proceedings of the Royal Irish Academy, 72 (B) (15): 253-274.

O'RIORDAIN, C. E., N.D., *A Catalogue of the Collection of Irish Marine Crustacea in the National Museum of Ireland.* Stationery Office, Dublin.

O'ROURKE, F. J., 1970. *The Fauna of Ireland.* Mercier Press, Cork.

PRAEGER, R. L., 1901. "Irish Topographical Botany". *Proc. R. Irish Acad.,* Dublin VII (Third Series) I-CLXXXVIII, 1-410.

PRAEGER, R. L., 1934. *The Botanist in Ireland.* Hodges, Figgis, Dublin. Republished EP Publishing, Wakefield, England.

QUICK, H. E., 1960. "British Slugs", *Bulletin of the British Museum* (Natural History), 6 (3) Zoological Series 103-226.

RUTTLEDGE, R. F., 1966. *Ireland's Birds.* Witherby, London.

SOUTH, R., N.D. *The Moths of the British Isles.* 2 Vols. Warne, London.

STEPHENS, D., 1973 (English edition). *Dolphins, Seals and Other Sea Mammals.* Collins, Glasgow.

TANSLEY, A. G., 1939, 1949 corrected and issued in 2 Vols. *The British Islands and their Vegetation.* Cambridge University Press.

TAYLOR, P., 1960. *British Ferns and Mosses* Eyre and Spottiswoode, London.

WAKEFIELD, E. M. and DENNIS, R. W. G., N.D. *Common British Fungi.* Warne, London.

WATSON, E. V., 1968 (2nd edition). *British Mosses and Liverworts.* Cambridge University Press.

WEBB, D. A., 1977. 6th revised edition (first published 1943). *An Irish Flora.* Dundalgan Press, Dundalk.

WENT, A. E. J., and KENNEDY, M., 1976 (3rd edition), 1957 (1st edition). *List of Irish Fishes.* Stationery Office, Dublin.

WHITEHEAD, G. K., 1960. *The Deer Stalking Grounds of Great Britain and Ireland.* Hollis and Carter, London. 1964, *The Deer of Great Britain and Ireland.* Routledge, & Kegan Paul, London. 1972, *The Wild Goats of Great Britain and Ireland.* David & Charles, Newton Abbot.

WITHERBY, H. F., JOURDAIN, F. C. R., TICEHURST, N. F., and TUCKER, B. W., 1938 and subsequent editions. *The Handbook of British Birds.* Vol. I pp 37-39. H. F. and G. Witherby, London.